HISTORY AND HISTORIANS

HISTORY AND HISTORIANS
A Historiographical Introduction

MARK T. GILDERHUS

Colorado State University

Prentice-Hall, Inc., Englewood Cliffs, New Jersey 07632

Library of Congress Cataloging-in-Publication Data

GILDERHUS, MARK T.
    History and historians.

    Includes bibliographies and index.
    1. Historiography. 2. History—Philosophy.
3. History—Methodology. I. Title.
D13.G395 1987   907′.2   86-18751
ISBN 0-13-390097-5

Editorial/production supervision and
    interior design: Barbara DeVries
Cover design: Ben Santora
Manufacturing buyer: Ray Keating

Printed in the United States of America

10 9 8 7 6 5 4 3 2

ISBN 0-13-390097-5   001

Prentice-Hall International (UK) Limited, *London*
Prentice-Hall of Australia Pty. Limited, *Sydney*
Prentice-Hall Canada Inc., *Toronto*
Prentice-Hall Hispanoamericana, S.A., *Mexico*
Prentice-Hall of India Private Limited, *New Delhi*
Prentice-Hall of Japan, Inc., *Tokyo*
Prentice-Hall of Southeast Asia Pte. Ltd., *Singapore*
Editora Prentice-Hall do Brasil, Ltda., *Rio de Janeiro*

to my daughters, Kirsten and Lesley,
and to the students in HY200

CONTENTS

PREFACE

This book provides a brief survey of Western historical thinking from ancient times until the present and an introduction to some of the main issues and problems in historiography, philosophy of history, and historical method. It seeks both to strike a balanced coverage of these issues and to make such concerns accessible to beginning students. Ordinary undergraduates in history need to encounter questions of theory in order to grasp the nature of the discipline but probably lack the prerequisites to take R. G. Collingwood straight. This small work may ease the transition.

I want to express thanks to my colleagues at Colorado State University who share an interest in exposing students to theoretical questions, notably George M. Dennison, Harry Rosenberg, and Manfred J. Enssle, and also to Jane Bowers and Steve Dalphin of Prentice-Hall who provided encouragement.

1

AIMS AND PURPOSES

Why bother with the study of history? What possible connections exist between an increasingly remote past and our own predicaments in the present? Can stories about other peoples in other places and other times have any meaning in an age of vaulting technology and traumatizing change? Is it reasonable to think that anyone can benefit from the experiences of others in a presumably unprecedented time when our political and economic systems falter, and the nuclear peril causes nightmares of dread? These questions hold more than rhetorical importance and compel serious answers. Undergraduates in all programs of study need to know what they can hope to learn and how their experiences will affect their capacity to think and act creatively in the future.

Skeptics have often argued that a knowledge of history will not provide much help. The American industrialist Henry Ford characterized history as "bunk." Although the observation probably tells more about the limitations of Ford's mind than about the nature of history, other luminaries have expressed similar reservations. In the seventeenth century, the French scientist and mathematician René Descartes worried that undue curiosity about the past would result in excessive ignorance of the present. Another Frenchman, François Marie Arouet de Voltaire, a philosopher and historian, described history as "a pack of tricks we play on the dead." Although he meant the comment as an appeal for history written more accurately, he inadvertently gave support to the cynical claim that historians invariably fall into one of three camps: those who lie, those who are mistaken, and those who simply do not know. Even so powerful a thinker as Georg Wilhelm Friedrich Hegel, a nineteenth-century German, feared that the only thing we can learn from history is that no one learns anything from history.

Undoubtedly the writing of history is a perilous venture. A common lament among historians is the fact that every day requires them to face up to their incomprehension of the world and their incapacity to interpret their evidence correctly. Surely they should rank among the humblest of people. Nevertheless, for many, the sheer joy of the endeavor makes the risk worthwhile. Some even have assigned to themselves important and useful functions. Most historians regard the study of history as a way for human beings to acquire self-knowledge. Edward Gibbon, the great English historian of the Roman Empire, sadly described the historical record as consisting of "the crimes, follies, and misfortunes of mankind." Though certainly indicative of a wretched and dismal state of affairs, his remark also held forth the possibility of escaping such conditions through rational inquiry. Transcendence over the past could come about only through knowledge.

Other historians have invoked their discipline as a kind of ethical sanction. Lord Acton, a Victorian Englishman, insisted upon maintaining "morality as the sole impartial criterion of men and things." He called upon historians to act as arbiters, defending the proper standards, out of an expectation that the threat of disapproval in the future might discourage incorrect behavior in the present. Historians should call malefactors to account for their misdeeds.

Still others presumed the existence of links between the

past and the future and suggested that comprehension of what had taken place might prepare for what will come about. How to get ready for the unknown has always posed a great problem. George Santayana, a Harvard philosopher, asserted early in the twentieth century that people who forget about the past are condemned to repeat it. This utilitarian conception saw in the discipline a way of developing workable strategies for survival. History comprised the recollections of all people. Santayana's claim affirmed that things learned from experience could aid in the avoidance of mistakes, pitfalls, and catastrophes in the future.

As a body of knowledge, history has a long and honorable tradition in Western civilization. Although definitions and points of emphasis have changed from time to time, the written narratives have always centered on human affairs and purportedly set forth truths. This latter claim means merely that historians have some good reason in the form of evidence for believing in the validity of their accounts. Paul Conkin, a contemporary American historian, has provided a succinct description. A history "is a true story about the human past."[1] Obviously the adjectives "true" and "human" are crucial. The quality of "truth" distinguished history from legend, fable, and myth, which admittedly may be "true" in some ways but usually not literally. The concern for the human past means that historians need to pay attention to the events in nature primarily when they affect the activities of people. Volcanic eruptions, for example, hold an interest mainly when they bury cities such as Pompeii.

From ancient times until the present, all peoples have told stories about themselves, their ancestors, and their origins. The Assyrians carved into stone monuments the names and deeds of their kings for everyone to see. The inscriptions also contained curses threatening to punish transgressors who violated the integrity of the record through defacement. The earliest tales usually dwelt upon extraordinary occurrences that were unusual, wonderful, fabulous, terrible, or miraculous. They told of spectacular events, often featuring supernatural powers, showing the gods and goddesses participating in the affairs of humans and oftentimes determining the outcomes of their acts. Some of the stories in Greek mythology, the Babylonian Gilgamesh epic, and the Hebrew Torah are examples. In the present day, such renditions sometimes confuse us because, by our standards, they seem to mingle the true with the untrue and the believable with the unbelievable. Nevertheless, they are not necessarily evidence of overwrought imaginations or low intelligence in ancient times. Rather, they bear out the historians' truism that different peoples in different times and different places literally saw and experienced the world differently. It may also be that very divergent conceptions of truth and believability have separated the present from the past.

"Real" history developed in the ancient world when iconoclasts announced their disbelief in traditional, oral accounts and insisted on setting the record straight. In Greece, early in the fifth century B.C., Herodotus of Halicarnassus composed some of the first "critical history" in our tradition by writing the "truth" about the Greek wars against the Persians. In putting together *The Histories,* Herodotus employed verifiable information, utilizing eyewitness accounts, some official records of state, and his own observations. His admirers regard him as "the father of history." Ever since Herodotus, historians have tried to tell true stories about the human past.

For two-and-one-half millennia, the study of history has satisfied many aims and purposes. In all likelihood, many students of the subject first acquired an interest out of simple fascination. As curious and inquisitive beings, they enjoyed the sheer fun of vicarious experience while asking "What was it like?" Through the exercise of imagination, they could take part in many things, the Punic Wars or the Renaissance. Some reveled in the possession of odd and esoteric pieces of information, possibly the kind of armaments used in battle during the Hundred Years' War or the lineage of Swedish kings, while others found in history a source of instruction, a way of making the course of human affairs intelligible, or at least some portions thereof. As noted by a European folksaying, no one is lost until they do not know where they have been. Historians seek to keep us from getting lost by locating us in time and figuring out where we have been.

Although simple curiosity is a sound reason for embarking

upon historical studies, professional scholars usually bring additional incentives to their work. Many are impelled by a strong sense of psychological necessity, a consideration which undoubtedly operates in most other areas of academic endeavor. Academicians want to bring some measure of order and predictability to the world. They dislike disorder and unpredictability because random and haphazard events defy comprehension and entail danger. Such vulnerability implies futility and the possibility of extinction. Scholars want to know what is likely to happen under various sets of circumstances. All academic disciplines strive to make accurate predictions about probable outcomes. Such is the case in physics, chemistry, sociology, and political science. It is also true in history, except that in this discipline the process takes place backward in time rather than forward. On the basis of fragmentary and imperfect evidence, historians make retroactive predictions about what probably happened in the past and, in so doing, they seek to define cause-and-effect relationships in order to make the flow of events understandable. Whenever historians make cause-and-effect statements—for example, Americans moved west because of the Panic of 1837—they affirm their belief in the intelligibility of events in the human world. Things happen for reasons, and inquiring minds can grasp them.

Such assumptions are deeply engrained in the traditions of Western civilization. Whether they are actually true is perhaps less important then the historians' conviction that they should be. For historians, the identification of cause-and-effect relationships establishes meaning and comprehensibility but can never be proven. They have to be taken on faith. As an example of an alternate view, Kurt Vonnegut's novel *Slaughterhouse Five* contains an intriguing vignette. The central character, the remarkable Billy Pilgrim, has the capacity to move around in time and space. He can travel into the past and into the future and also beyond the confines of Earth. In one episode, he is kidnaped by extraterrestrial beings from the planet Tralfamador. They put Billy Pilgrim on public display, locked up in a transparent geodesic dome with another captive, Montana Wildhack, a voluptuous movie starlet. The Tralfamadorans enjoy watching the two cavort about and also engage them in philosophical discussions. Billy Pilgrim amuses and awes his captors by affirming his belief that cause-and-effect relationships govern the course of events. Things happen because other things make them happen. The Tralfamadorans have a different notion. For them, things happen merely because they happen—randomly, haphazardly, inexplicably, chaotically. The adoption of any such world view would make the work of historians next to impossible.

History also provides a way to study the identity of people, both individually and collectively. In some ways, this function of history parallels psychiatry, in that both fields endeavor to clarify human behavior in the present by making knowledge of the past both conscious and explicit. Just as psychiatrists seek to treat aberrant or disturbed conduct by scrutinizing repressed or subconscious memories, so historians try to arrive at a fuller understanding of the actions of people by examining their history. Robin G. Collingwood, a British philosopher and historian, liked to suggest that human beings possess no nature; they have merely history. As malleable creatures, they become whatever their experiences make of them.

Even allowing for some exaggeration, Collingwood has a compelling point. Historical experience shapes and molds the identity of people in important ways. Most of us recognize this claim as a fact in our rituals. For centuries, Jewish people in their Passover feasts have told the story of ancient Israel and the special covenant with Yahweh, their God, and have managed, in spite of isolation and dispersion, to maintain a collective sense of group identity. In a less profound way, Fourth of July ceremonies in the United States serve a similar purpose. By invoking patriotic lore about the American Revolution, the people of the country establish a sense of solidarity by celebrating the origins of the nation. Just as a single person might explore the question "Who am I?" by thinking through life experiences, historians tell the life stories of whole peoples. When we ask, "Who are the Persians?" or "the Germans?" or "the Taiwanese?" or "Who are we?" the narratives of history provide one place to begin.

Another reason for studying history is utilitarian and practical. According to this rationale, history has a useful application because it helps us better to calculate the anticipated consequences of our own acts. George Santayana probably had this idea in mind when he said that people have to repeat the past if they forget it. His words should not be taken too literally. The Second World War will not happen again, even if we neglect to read and write about it. Santayana meant something deeper. He knew that history is the collective memory of humankind and that the onset of a mass amnesia would have bad effects. For one thing, it would prevent the young from learning from the old. Each generation would have to find fire and invent the wheel over and over again. Without memory, we would have trouble functioning and making do in the world. What each generation transmits to the next can be understood in some measure as lessons in the art of survival.

Karl R. Popper, a philosopher, pointed out another facet of the problem. He believed that, above all, social scientists and historians should contemplate the unintended consequences of deliberate human acts. Sometime things go wrong. Historical actors set out to accomplish a set of goals and actually bring about unanticipated or contrary results. Popper wanted students of human affairs to investigate the linkages between intentions and outcomes. Napoleon's quest for domination in Europe destroyed feudal structures and cleared the way for modernization. In South Vietnam, the United States employed military force in defense of the right of self-determination and facilitated the obliteration of a small country. Such ironies, sometimes tragic, sometimes comic, abound in human experience. The theologian, philosopher, and historian Reinhold Niebuhr pondered this maddening issue in his book *The Irony of American History* and warned that the actual consequences of our acts sometimes subvert our commitments to high ideals. We need to be careful in pursuing grandiose purposes because so often they go awry. If we could better reckon the relationship between aims and outcomes, our chances of behaving more creatively in the world would improve.

Historians typically construct their narratives by affirming the existence of cause-and-effect relationships and appraising the connections between the actions of historical figures, their motives, and the consequences. The following three-stage model has merely descriptive, not prescriptive, intent. It shows how many historians do carry out their work but sets forth no requirement that they must proceed in this fashion.

During the first step, historians begin their inquiry by asking, What happened? How did the historical actors behave? What did they do? This part is the easiest. As long as some kind of artifacts exist, possibly oral traditions or stone tablets or manuscripts or diaries or newspaper accounts or official records of state, historians can arrive at some determinations. If no remnants exist at all, then no written history is possible.

During the second step, historians must account for the actors' behavior by asking the question, Why? The answer usually calls for an explanation or interpretation and has tricky implications because it entails an assortment of methodological and theoretical dangers. Historians traditionally have employed a "rational human" model of behavior in framing their explanations. They have assumed that most people set rational goals for themselves and then seek to achieve them through the exercise of reason and logic. In more recent times, dissenters have criticized this approach as hopelessly antiquated. Marxist scholars, for example, have argued that economic and class relationships determine behavior and that ostensibly rational processes serve to mask hidden purposes. Similarly, the advocates of psychohistory have rejected the "rational human" model. In their efforts to apply psychoanalytical theories to history, they find the wellspring of human behavior not in reason but in repressed impulses tucked away deep within the recesses of the psyche. Such disparities of understanding mean that discussions of motive are always tentative and uncertain.

In the final step, many historians try to scrutinize the consequences of a course of events and to arrive at an evaluation. How did things turn out, for good or for ill? Who benefited and who suffered? Did the outcome make the effort worthwhile? On such big and significant questions, historians seldom agree. The reason is obvious. Any attempt to address them will bring differing and rival value systems into play, and no means exist by which to reconcile the ensuing judgments. How can historians accurately measure the costs and gains of the Mexican Revolution after 1910? Would "natural progress" have made the Mexicans better off if the upheaval had never occurred? No scholar has any good way of knowing for sure.

Historians thus practice their craft in a kind of intellectual minefield in which all sorts of unknown and unanticipated dangers pose threats, and the best of projects can blow up before them. The evidence invariably is too sparse to tell the whole story. Even when it exists in abundance, the difficulties of explanation, interpretation, and evaluation are immense. Yet, historians persist and endure in their toil, seeking to render some small portions of human experience intelligible. This book is intended as an introduction to the practices and patterns of historical thinking.

RECOMMENDED READINGS

Engaging observations and pungent definitions of the nature of history appear in Ferenc M. Szasz, "The Many Meanings of History," *The History Teacher*, VII (Aug. 1974), pp. 552–63; VII (Nov. 1974), pp. 54–63; VIII (Feb. 1975), pp. 208–16; and another piece provided by subscribers, IX (Feb. 1976), pp. 217–27. Introductory works considering aims and purposes include Fritz Stern (ed.), *The Varieties of History, From Voltaire to the Present* (New York: William Collins Publishers, Inc., 1956); Allan J. Lichtman and Valerie French, *Historians and the Living Past, The Theory and Practice of Historical Study* (Arlington Heights, IL: AHM Publishing Corp., 1978); Carl G. Gustavson, *The Mansion of History* (New York: McGraw-Hill Book Co., 1976); and Arthur Marwick, *The Nature of History* (New York: Dell Publishing Co., 1970). R.G. Collingwood, *The Idea of History* (New York: Oxford University Press, 1956), although difficult for beginners, is indispensable. Karl R. Popper's discussion of "Prediction and Prophecy in the Social Sciences" appears in the collection edited by Patrick Gardiner, *Theories of History* (New York: The Free Press, 1959), pp. 276–85.

2

THE BEGINNINGS OF HISTORICAL CONSCIOUSNESS

The traditions of Western civilization possess a distinctive level of historical consciousness. This awareness developed in large measure from the legacies of the Jews, the Greeks, and the early Christians. The Jews and later the Christians imposed upon the events of the past a sense of meaning, structure, and process. For them, the past merged with the present and the future and moved inexorably toward a set of definite and knowable goals. The Greeks, meanwhile, contributed an insistence upon studying the past critically and scientifically in order to determine the truth. For them, the distinction between history and mythology became fundamental.

In contrast, the earliest human beings had little historical consciousness. They lived in an expansive present in which the urgencies of mere survival pressed incessantly upon them. Indeed, the terrors of the past—the recollection of impermanence, hunger, death, catastrophe, and destruction—may have created psychological barriers against the act of remembering. The trauma of mere existence removed all meaning from past events, except perhaps the negative connotations. To whatever extent the earliest people conceived of a time dimension, their impressions took on a cyclical form—that is, a conception of events moving more or less meaninglessly in a circle. Things occurred, went away, and then recurred, much in the fashion of days, nights, and seasons. For ancient peoples, the familiar and predictable patterns of nature became a way of organizing the unfamiliar and unpredictable happenings in the human world.

Religious myths, legends, and fables satisfied the need of ancient peoples to know about their origins. These tales often posited the existence of a "golden age" during which the people lived harmoniously among themselves and with nature. The story in Genesis about Adam and Eve in the Garden of Eden is an example. Such accounts usually presumed a close relationship with the gods and goddesses in which the latter often possessed humanlike characteristics and achieved their purposes through the manipulation and control of the lesser beings. In Greek mythology, the gods competed with one another for love and power, waged war on one another, and sometimes failed. Infallibility was not necessarily an attribute of supernatural beings in the ancient world. The deities also had a capacity for cruelty when they sought vengeance and inflicted retribution. When disasters and cataclysms took place, ancient people normally attributed them to the will of the supernatural powers. The Greek notion of hubris, meaning self-destructive pride, required that the gods bring down mere mortals who overreached themselves. In all likelihood, such myths and legends also had political importance. By affirming a connection with the divinities through lineage or proximity, ancient rulers could establish the legitimacy of their right to govern.

Some ancient peoples kept no records and hence have no recoverable history. Others, such as the Egyptians, Sumerians, Assyrians, and Hittites, left written artifacts dating from the third and second millennia B.C. Many of these consisted of lists and inscriptions recounting the deeds of great men. They testified to the emergence of a primitive historical consciousness, perhaps a sense of chronology, but showed little appreciation for the effect of one event upon another or the interrelationships among them. Instead, these registers presented bits of information in isolation from others. Sometimes they recited the major accomplishments or occurrences during the reign of a king, noting the year the wall was built or the temple completed, the time of the great famine or plague, or when enemies laid waste to the city. Accounts of warfare figured prominently. Graphic, brutal, even sadistic, descriptions reported terrible punishments visited upon conquered foes. An Assyrian document reported these acts performed by a king named Assurnâsirpal:

> 600 of their warriors I put to the sword; 3,000 captives I burned with fire; I did not leave a single one of them alive to serve as a hostage ... Hulai, their governor, I flayed and his skin I spread upon the wall of the city; the city I destroyed, I devastated, I burned with fire.[1]

In another episode, the same monarch cut off the hands, noses, and ears of another set of adversaries, put out their eyes, and removed their heads, from which he constructed a pillar. Such documents contained the raw material of history but did not in themselves constitute history. They provided no interpretations or analyses. They reported victories in battle but not the means of winning. The bloodlust exhibited in them suggests why the horror of recollecting the past might have repelled ancient peoples.

     The Jews of ancient Israel developed a very different outlook. For them, history became more important than for any other ancient people. Indeed, it became a kind of obsession, the comprehension of which established the meaning of their existence and affirmed their own unique sense of destiny. Through the processes of history, the ancient Hebrews forged a distinctive relationship with their God, Yahweh. He had chosen them as a special people with particular obligations and responsibilities. The experience of deliverance from bondage in Egypt into the Promised Land became a memory seared upon the Jewish consciousness. In return, they had a commitment to comport themselves according to the Law. By their conception of the Covenant, the Jewish people would enjoy peace and plenty as long as they complied with the divine will, but they would suffer calamity and woe if they violated the Commandments. The Jews saw the hand of God in human affairs. He rewarded good and faithful behavior and punished transgressions. The Jews reported their history accordingly in the books of the Old Testament.

Hebrew historical writing was more the product of religious experience and faith than a manifestation of critical or rational inquiry. The Jews interpreted the events in the lives of their people according to intense convictions. Bias and inconsistency, to be sure, crept into their narratives. For example, Jewish writers sometimes incorporated different versions of the same events from diverse oral traditions. Nevertheless, they also displayed a capacity for hardheaded objectivity. A remarkable passage in the second book of Samuel describes the antics of King David at a festival, "leaping and dancing" about, even uncovering himself before the handmaidens of his servants. Much disturbed by the scene, Michal, the daughter of Saul, felt disdain for David and later, upon encountering him, remarked disparagingly, "How glorious was the king of Israel today." As Michael Grant, a historian of the ancient world observed, "David does not show up very brilliantly there."[2] In spite of his heroic stature, he was depicted unidealistically. The Jews knew how to exercise some measure of detachment.

The Greeks contributed something of immense significance in the development of historical thought. They invented critical history as a method of sorting out the true from the false. In the classical Greek language, the term *histor* referred to a learned man who settled legal disputes. He looked into the facts and determined the accuracy. Subsequently as Michael Grant explained, the word *historie* meant "a search for the rational explanation and understanding of phenomena." In the fifth century B.C., two geniuses, Herodotus and Thucydides, brought about an intellectual revolution by employing such techniques and creating the writing of history.

In some ways, it is startling that such a development should have taken place in Greece, for the Greeks were not especially historically minded. To the extent that they conceived of a time dimension, they embraced a cyclical mode of thinking. Homeric poetry, such as *The Iliad* and *The Odyssey*, recounted tales of earlier times, but they were not history, even though the Greeks may have thought of them as such. Rather, they consisted of legends, myths, and fables in which gods, goddesses, and heroes acted together, and supernatural forces accounted for the progression of events. Divine wills figured prominently in human affairs. Before history could exist, a test of reason had to take place. The great accomplishments of Herodotus and Thucydides provided the model for all subsequent written history. Many modern scholars still regard them as ranking among the greatest historians of all time. Historiography, which is the history of historical writing, holds them in high esteem.

The Histories by Herodotus of Halicarnassus told of the Greek wars against Persia during the third decade of the fifth century B.C. The first portion of the work provided historical background, exploring the origin of the quarrel between Greece and Persia. The remainder recounted the details of Persian expeditions against Greece under the kings Darius and Xerxes. Little is known about Herodotus' life. Presumably he lived about a generation after these events, and he traveled widely in order to collect material. The Persian Wars impressed him as the most important events in world history, and he made them the framework of his narrative. As he explained, he had two purposes in mind: "to preserve the memory of the past by putting on record the astonishing achievements both of our own and of the Asiatic peoples" and "to show how the two races came into conflict."[3] He achieved these aims and much more. Frequent digressions in the narrative dispensed an array of fascinating and odd information. Herodotus wanted to know why the Nile flooded each year and how the pyramids came into being. Many things piqued his curiosity, and intriguing observations followed. The Babylonians, for example, lacked medical doctors and made it a practice to put invalids in the street so they could receive advice from passersby, and the Persians made it a habit to deliberate twice on important issues, once while drunk and once while sober. The one condition served as a check on the other.

Herodotus liked a good story but also employed rigorous methods. He checked his information against the reports of eyewitnesses and participants and also consulted the documents available to him—inscriptional records, archives, and official chronicles. He also departed from the custom of explaining human events as the outcome of divine will. To be sure, he never succeeded in rendering a completely secular account of history. The deities still had a role to play in human affairs. But Herodotus, more than any predecessor, interpreted the course of events as the product of human wills.

Thucydides, an Athenian, wrote *The Peloponnesian War*, an account of the struggle between Athens and Sparta during the last three decades of the fifth century B.C. Thucydides took part in the conflict as a military commander and suffered the effects of the great plague in Athens, a horror which he vividly described in his writing. When the humiliation of a military defeat forced him into exile, he spent the next twenty years gathering materials for his work. Like Herodotus, he picked warfare as his subject. As he explained, he began his history "at the very outbreak of the war, in the belief that it was going to be a great war and more worth writing about than any of those which had happened in the past." He also wanted his history to be useful. Expressing a view typical of his time, he expected that "what happened in the past . . . will, in due course, tend to be repeated with some degree of similarity."[4] Thucydides intended his writing to have instructional importance as a guide to action in the future. Although history would never repeat itself exactly, he anticipated the development of parallel circumstances and believed that the consciousness of history would bestow many benefits. All leaders should learn from the mistakes of the past. Indeed, they could master the arts of politics, statecraft, and warfare only from the study of history.

Thucydides was scrupulous in his methods. As a careful student of causation, he took pains to identify the process by which the Peloponnesian War came about. His analysis attributed the real or underlying reason to Sparta's fear of Athens' growing power. He also paid heed to individual motives but used a convention very troublesome for modern scholars. Like Herodotus, Thucydides had historical figures deliver speeches in which they revealed their aims and intentions. Critics have attacked such monologues as false. Even if the historian had heard them, how could he remember the details? Apologists, in contrast, have pointed out that while not literally true, such devices enabled Thucydides to elucidate his understanding of the actor's character and goals. In a sense, Thucydides had them say what the situation called for.

Thucydides, more than Herodotus, explained events in secular terms. In *The Peloponnesian War*, things happened not because the gods willed them but because of human activities. People had to endure the consequences of their own acts. Thucydides also strove for objectivity. His detachment and single-minded determination to stick to the essentials of the story made for less entertainment than in Herodotus. No amusing digressions provided diversions for the reader. But the undoubted brilliance of his work inspired David Hume many centuries later to write, "The first page of Thucydides is the beginning of all true history."[5]

The quality of Greek historical writing declined after Herodotus and Thucydides. Much became lost to posterity. Nevertheless, the two masters exerted strong influence over their progeny. In the second century B.C., Polybius, a Greek, embraced them as his models. He insisted that a historian must travel to see the sites, participate in the events, and utilize the documentary records. He must also have a grand theme. Polybius's *Universal History* told of Rome's expansion over the whole of the Mediterranean world. He also intended that his work have useful effects. As a form of practical instruction, it should equip its readers better to act upon the future. Polybius passed on to the Romans the best of the Greek traditions in historical writing.

The Romans, strongly affected by the Greeks, also concentrated their attention on the political and military activities of the ruling elites. Late in the first century B.C., Sallust wrote from an antiestablishment point of view about the affairs of state, taking the leaders to task for their mistakes and misdeeds. Titus Livy reported upon the reign of Augustus, a period of forty-five years during which peace returned after an era of conflict and civil war. His *History of Rome* comprised an astounding 142 volumes. Unlike the historians who were also men of action, notably Thucydides, Titus Livy devoted his whole life to research and writing. The Romans also developed an interest in biography. Among the most famous, Plutarch, writing in Greek, studied the lives of soldiers and statesmen, and Suetonius composed the *Lives of the Caesars*.

Two Romans, Julius Caesar, and Cornelius Tacitus, had special significance. Caesar's *Commentaries* recounted firsthand his military exploits against the Gauls, Germans, and Britons in the middle of the first century. This unique and original work, spare and unadorned, established a stylistic model of Latin prose and also a new literary genre. As the memoirs of a field commander, Caesar's writings showed less interest in character and personality than in action. As a result, future scholars obtained more information about the Gallic War than any other military operations in the ancient world. Caesar also wrote with an eye toward his reputation before the governing classes in Rome and before posterity. Political and military memoirs are seldom self-effacing.

Tacitus, one of Rome's greatest historians, produced several works on political and military subjects. The most notable, *The Annals of Imperial Rome*, described the affairs of the empire from the beginning of Tiberius' reign in 14 A.D. until the year 68, soon after the death of Nero. Written early in the second century A.D., the book vividly depicted personalities and also the corruption and degeneracy of Rome's rulers. Although Tacitus professed scholarly detachment and objectivity, the excesses of the emperors and the cliques around them outraged his moral sense, and he called them to account, often with delicious, sardonic humor. For Tacitus, the city of Rome held great importance as the seat of power and the center of the world. He used the traditional format, a narration of events, year by year, recounting the rise of autocracy and despotism during the first century. Tacitus possessed little faith in the goodness of human nature. For him, the concentration of political power in one man always entailed hazards, chiefly by making it possible for character flaws to have maximum destructive effects. Tacitus' tales of wrongdoing in the struggle for personal advancement featured adultery, incest, murder, matricide, patricide, and infanticide. He shared with Lord Acton a view expressed some eighteen centuries later, that "power tends to corrupt, and absolute power corrupts absolutely." The careers of the debauched Tiberius, the eccentric Caligula, and the brutal Nero provided cases in point.

Meanwhile, an emerging Christian view of history took on more cosmopolitan forms, extending beyond the provincial and nationalistic fixations of Greece and Rome. It developed slowly during several centuries and out of harsh experience. The earliest Christians had little historical consciousness. They conceived of themselves as living in a small, closed universe, very near the conclusion of time. Since the end of the world and the Second Coming dominated their expectations, the future held more significance than the past. Nevertheless, they felt compelled to confront historical problems.

For one reason, as they undertook missionary work into the world of the Gentiles, they needed to establish the believability of their claims. From various sources, oral traditions and the like, they compiled the Gospels, consisting of stories about the life and sayings of Jesus, his ministry, crucifixion, and resurrection. The Christian insistence that such events actually had taken place grounded their beliefs in an interpretation of historical experience and established the veracity of their claims before believers. History also became important when the early Christians tried to establish proper connections with the Old Testament. The identification of Jesus with the Messiah of Hebrew prophecy necessarily engaged them with the teachings of the Hebrew prophets. Once the early Christians decided to retain the Old Testament as Holy Scripture, they had to formulate an explanation of the proper relationship between the past, the present, and the future. Ultimately they conceived of the whole of the Old Testament as an anticipation of Christ and preparation for him.

Christianity later took on some of the attributes of universal history through its efforts to appeal to pagans. To obtain prestige within the Roman Empire, Christians thought it necessary to show that Hebrew wisdom had roots more ancient than Greek philosophy and that Moses came well before Plato. The attempt to link the experiences of Jews and Gentiles compelled Christians to develop ideas about the secular history of all humankind. The very nature of the Scriptures, of course, also contributed by starting with the Creation and then telling about the beginnings of the human race and the division into different languages and nations. Indeed, the book of Genesis established a pattern copied by the early chapters of later, universal histories until the eighteenth and nineteenth centuries.

The early Christians suffered oppression and persecution for their beliefs but finally captured the Roman government through the conversion of the Emperor Constantine early in the fourth century. This triumph encouraged and made possible the writing of ecclesiastical history, detailing the rise of Christianity throughout the Mediterranean world. Eusebius of Caesarea in Syria Palestina was an early practitioner. His *History of the Church* and other works demonstrated how all previous history, especially the Hebrew experience, had led to the Christian revelation. Eusebius took pains to rebut the pagan charge that the newly invented Christian religion appealed only to ignorant people. Quite the contrary, he argued, the traditions stretched backward in time into a glorious past far beyond the Greeks. His work celebrated the Christian victory in taking possession of Rome.

A century later, St. Augustine, the bishop of Hippo in North Africa, wrote *The City of God* under very different circumstances. In the year 410, Alaric and the Goths captured the city of Rome for three days and sacked it. This catastrophe, verging on the unimaginable, ignited fears that resurgent barbarism would overwhelm civilization and also provoked charges from pagans that the responsibility resided with Christians. By this way of thinking, the conquest of the Eternal City signified retribution for the neglect of the ancient Roman gods. In response, Augustine wrote his powerful work between 413 and 426. To an extent, he labored as a partisan and polemicist in defense of the Christians. In the first ten chapters, he absolved them of blame, showing that equivalent disasters had taken place before, even when the pagan gods supposedly held sway. In the remaining ten chapters, he developed a new and unique approach to universal history. Drawing in some measure upon earlier Jewish and Christian writings, he developed an analysis that would dominate European thinking for over a thousand years.

As an organizing principle, Augustine established a dualism, distinguishing fundamentally between the earthly "city of man" and the heavenly "city of God." Though intermingled on earth for the remainder of time, the two communities took on distinctively different characteristics. The first, the secular city, featured self-love and the ways of the flesh. The second, the eternal city, dedicated spiritual love to God and transcended mere mortality. Although Augustine traced the progess of both cities until his own time, his main preoccupation centered on sacred history, that is, his attempt to elucidate the ways of God toward his creation and to establish the appropriate correlations of events with Biblical accounts.

Augustine's work also had important implications for the philosophy of history. His sense of time, derived from Hebrew conceptions, rejected outright the Greek idea of cyclical movements. For Augustine, endless revolvings and pointless repetitions would have rendered history meaningless; in effect, a nullification of divine influence and purpose. Rather, he thought of history as moving along a line with a clear beginning, marked by the Creation, a middle, and an end. The birth and death of Christ denoted the central events, and the salvation of all believers at the termination signified the completion of the process. The concluding triumph of the city of God over the city of man would result in the fulfillment of the final aim, the transcendence of believers beyond history into the realm of the eternal.

Augustine's conviction, endorsed in art, literature, and music by Michelangelo, Dante, and Handel, made profound imprints during the Middle Ages. The schema, characterized by the two cities and the movement of events along a line, inspired Christian writers through the medieval period and after. For them, teleology and eschatology —the attempt to comprehend the final causes and the last days—held prime significance, and they based their predictions about the end of time on an Augustinian understanding of God's plan, working out his will toward the world in and through history.

For over ten centuries, writers on historical subjects during the Middle Ages labored under Augustine's influence. Though Europeans during this time made distinctions between secular and sacred history, they paid the greatest attention to the latter and composed ecclesiastical or church histories, designed primarily to delineate the progress of God's work in the world. Characteristically the writing took the form of annals and chronicles, set forth by priests and monks. The forms became stylized, based on Augustine's model. Normally they contained a resume of universal history since the Creation, influenced often by Old Testament accounts, an affirmation of God's purpose and will, and then statements of occurrences in more recent times, showing the immanence of the divine presence in reality.

The annals, the more rudimentary, consisted of lists in brief entry, a record of memorable events in a locality during a single year in some kind of yearbook or register. An aggregation of such entries, taken together and arranged in an annual sequence with some additional narrative, produced a chronicle. Though usually geared to record God's activities and manifestations, such sources also contained a great deal of information about secular life in Christian communities. The compilers told about the goings on within the monastery, the deaths of bishops and kings, the outcomes of battles, and the construction of new churches. They also reported upon unusual episodes, often taken as omens or portents from which to gain foretellings of the future. Eclipses and floods, the appearance of strange animals, or the birth of malformed children provided occasions for uncertainty and wonderment.

Though annals and chronicles became the main vehicles for historical writing during the medieval period, the priests and monks usually assembled their records in anonymity and hence posed a problem for modern scholars who seek to identify them. Some carried out their labors for purely personal reasons, others to keep an official record for some kind of patron, possibly the commission of a royal court. The problem of determining authorship became even more difficult because of the prevailing practice, sometimes over several generations, of circulating the documents from one monastery to another for emendations and exchanges of information.

Although in many instances they were the work of several authors, the annals and chronicles bore some uniform characteristics. For example, they testified to a near universal belief in divine providence. For medieval writers, one God with paternal authority indisputably stood above humanity, watched over the course of events, and regularly influenced them through divine intervention. Such expressions of faith, of course, influenced the discussions of human behavior and the analyses of events. The compilers of annals and chronicles looked upon religion as the first concern and believed too that history moved teleologically toward a foreordained conclusion directed by God.

Such proclivities produced a tendency to moralize. Medieval writing often rendered verdicts and judgments, raising questions about their objectivity and truthfulness. Contemporary scholars have to recognize the issue as a modern one. Most medieval chroniclers would not have seen a problem. They looked upon an intention to tell the truth as the main responsibility. Procopius, for example, declared that "truth alone is appropriate to history," and Matthew Paris invoked an even stronger sanction against historians. "If they commit to writing that which is false, they are not acceptable to the Lord." To assure credibility, medieval writers usually informed readers of their sources of information. The following examples suggest some additional characteristics.

Procopius, a Byzantine of the sixth century, composed contemporary histories of the wars of the Emperor Justinian I against the Ostrogoths, Vandals, Visigoths, and Persians. His major work is known as the *History of the Wars*. As a writer well versed in classical Greek historiography, he embraced as his models the works of Herodotus, Thucydides, and Polybius. He also emulated their methods. He used his political connections to good advantage and managed to accompany the generals onto the battlefields. Probably he also gained access to state papers and official archives. He merits a high place among the ranks of medieval historians because of the quality and persuasiveness of his writing. Nevertheless, modern authorities concede that they have no way to test his trustworthiness, since no other sources exist against which to measure his accounts.

The Venerable Bede engaged in a very different kind of scholarship early in the eighth century. He spent most of his life in the monastery at Jarrow in Northumbria in the north of England, studying, teaching, and writing on religious, educational, and historical subjects. One of the great centers of learning in its day, the monastery of Jarrow had an excellent library. Modern scholars acclaim Bede as one of the most learned men of the early Middle Ages. His *Ecclesiastical History of the English People* is the first reliable description of early Britain, the sacred and secular aspects. Bede initially became interested in history while pondering the problem of establishing an accurate chronology of events and the proper dates for Christian festivals. Later he became intrigued with the arrival of Christianity in England, a singular and striking thing, the expansion of the Church through the country, and the effects upon the cultural and intellectual life of the country. He discussed extensively the literary activities of monks. Though Bede seldom conducted research in the archives and repositories of the great churches and abbeys and never participated in battles, he had a high regard for valid sources and often arranged to have written accounts and other narratives brought to him from Canterbury.

In the twelfth century, Otto of Freising emerged as the greatest of the German chroniclers. Born into a noble family, a nephew of the emperor, Frederick I Barbarossa, Otto became the bishop of Freising, near Salzburg in Austria, and stayed there all his life. He wrote two important works, *The Deeds of Emperor Frederick Barbarossa*, a laudatory political biography of his uncle, and the *Chronicle or History of the Two Cities*, written as a universal history and modeled on the work of St. Augustine. Otto of Freising employed a linear conception of history and presumed the existence of a beginning, the Creation, and an end, the Second Coming, after which Christians would enjoy everlasting bliss in eternity. He vividly presented the notion of the two cities and took it more literally than St. Augustine, seeing the Church actually as the representative of the Holy City. Yet, as a member of the royal family and a close observer of imperial affairs, he was somewhat more conciliatory in his attitude toward the secular state. He possessed little faith that Christianity could moderate the baser aspects of human nature in this world.

Two chroniclers of the late medieval period, Matthew Paris and Jean Froissart, also merit mention. Matthew Paris, an English monk in the thirteenth century, recorded the affairs of church and state in his own time. His *Greater Chronicle* incorporated an edited version of *Flowers of History*, the work of a predecessor, Roger of Wendover, but Matthew's narrative upon reaching the year 1236 became an independent and original source. More broadly gauged than much medieval writing, it described events in England and elsewhere, including western Europe, the Papal states, the Latin Empire of Constantinople, the Holy Land, and even Russia. He paid special heed to politics and international intrigue but also liked to remind his readers of God's presence, divine providence, and death and judgment.

In the fourteenth century, the *Chronicles* of Jean Froissart, a French priest, also centered on political and military affairs of state. He wrote with self-conscious detachment and endeavored to tell the truth in an age of chivalry. He stirringly depicted battles and heroic deeds by knights in combat. He also assumed a divine role in human affairs but worked hard to obtain accurate information. As a historian preeminently of the aristocracy, he took little interest in the mundane lives of the lower classes.

The historiographical traditions of the medieval period thus modified the Greek approach by putting supernatural powers back into history. They persisted well into the modern era; but, by the fourteenth century, such writings had become formalistic and repetitive. Medieval chroniclers typically concentrated upon affairs in their own time and relied upon earlier authorities for information about previous ages. By and large, they lacked the intellectual and methodological means to employ original sources and to recover their meaning and significance. Instead, they had to depend upon the veracity of their predecessors. In large measure for this reason, history as a discipline had little standing in medieval universities. No one could demonstrate the truthfulness and believability of its claims. But important changes took place between the ends of the fourteenth and nineteenth centuries. During those five hundred years of turmoil and upheaval, history achieved such status and obtained new capacities to verify its principal claims.

RECOMMENDED READINGS

Historians at Work, Vol. 1: *From Herodotus to Froissart*, edited by Peter Gay and Gerald J. Cavanaugh (New York: Harper & Row, Publishers, 1972) provides a convenient anthology of the writings of the great historians in the ancient and medieval worlds. *The Ancient Historians* by Michael Grant (New York: Charles Scribner's Sons, 1970) considers the Greeks and the Romans. Herbert Butterfield, *The Origins of History* (New York: Basic Books, Inc., Publishers, 1981) describes the inception of historical thinking and Christian renditions. Contributions in the medieval period are considered in Joseph Dahmus, *Seven Medieval Historians* (Chicago: Nelson-Hall, 1982) and Indrikis Sterns, *The Greater Medieval Historians, An Interpretation and a Bibliography* (Washington, D.C.: University Press of America, 1981). John Barker, *The Superhistorians, Makers of Our Past* (New York: Charles Scribner's Sons, 1982) contains fine essays on Herodotus, Thucydides, and St. Augustine. A sophisticated and detailed study by Ernst Breisach, *Historiography, Ancient, Medieval, & Modern* (Chicago: The University of Chicago Press, 1983) covers in much greater depth many of the issues under consideration here.

3

HISTORICAL CONSCIOUSNESS IN THE MODERN AGE

The historical consciousness of the modern age developed gradually during the course of five hundred years between the ends of the fourteenth and nineteenth centuries. The change entailed a shift away from reliance on supernatural explanations and toward the development of secular modes of explanation. As the historians came to realize, it was one thing to say that God acted in history and another to determine with any precision just where and when. History acquired a more "scientific" outlook, at least insofar as the methods of research, criticism, and analysis became more vigorous and acute and as the practitioners strove to keep abreast of stunning advances in other areas of human knowledge.

In the modern age, history, much as other forms of academic inquiry, abandoned attempts to determine ultimate or final causes. Upon reflection, the reasons become immediately obvious. They have to do with the limitations of research methodology. No modern scholar working in a reputable discipline can claim legitimately to possess the means to verify any claims set forth about the role of divine influences in the physical and historical worlds. Any such claims amount to a manifestation of faith. Rather than run such risks in their capacities as professional scholars, historians have concentrated their efforts on devising methods by which to expand understanding of the knowable world.

To such ends, they have sought ways of enlarging in the present our comprehension of all kinds of artifacts from the past. In this sense, they tried to recapture the spirit of the Greeks and to approach the study of history critically, putting more demands upon the evidence than upon the traditions of faith and authority. The transformation took place incrementally over a long period of time, beginning with the resurgence of interest in the ancient world during the Renaissance, reacting later to religious, scientific, and philosophical currents, and culminating with the establishment of university-based historical studies in the nineteenth century. During this so-called "golden age," Leopold von Ranke and other scholars transformed the practice of history into a profession and laid down the main directions of development until the present day.

Among the first intimations of an impending break with medieval historiography during the early stages of the Renaissance, the life labors of Francesco Petrarca, usually anglicized as Petrarch, aimed at the recovery of the traditions of ancient Rome. A native of the city of Florence, Petrarch devoted his attention to the preservation and transcription of classical writings. For him, the Roman experience incorporated the best attributes of humankind. "What else, then," he asked, "is all history, if not the praise of Rome?" Though concerned primarily with the collection of literary artifacts, he also wrote a history of Rome, *Lives of Illustrious Men*, and he contributed significantly to the establishment of a new, secular interpretation, but he did so without directly attacking Christian authority. His search for Rome set forth an alternative vision of human beings and their world in which real events had more than just symbolic importance. The actuality of human striving and attainment for him held great allure.

In the sixteenth century, another Florentine, Niccolò

Machiavelli, focused attention on the human dimension in history through his writings on politics. Born in 1469 as the son of a lawyer, Machiavelli's education included readings in the Latin and Italian classics. Beginning in 1494, he embarked upon a political career when the combined effects of a French invasion and Savonarola's religious uprising resulted in the ouster of Lorenzo "the Magnificent," the ruler of Florence and a member of the house of Medici. Machiavelli subsequently served the Florentine republic in various capacities at home and in other states until the restoration of the Medicis in 1512. Forced into exile on a farm outside of the city, Machiavelli then spent his time trying to regain his lost influence and writing about the conduct of politics, the true description of which became his obsession. His most famous work, *The Prince*, set forth a body of precepts, showing that the actual methods of governance had little to do with medieval theory. According to this handbook on the uses of political power, the wise prince had to know when to employ duplicitous or cynical means. Power and the proper utilization of it to achieve self-interested advantages constituted the great theme. History for Machiavelli turned into a kind of grab bag of examples by which to illustrate his maxims. His *History of Florence* detailed the intrigue and machination so characteristic of the politics of his own city and depicted actual behavior as a function of opportunism and aggrandizement.

The rediscovered wisdom of the Greeks and Romans inspired the Italians and served them as models, e.g., the writing of Thucydides, Polybius, and Titus Livy. Following such leads, the ensuing narratives often dwelt upon political and military themes. Machiavelli's younger contemporary, Francesco Guicciardini, adhered to the classical themes in his *History of Italy*. A better historian than Machiavelli, Guicciardini carried out more research, notably in the papal archives, understood more subtly the diversity of human nature, and espoused a less provincial outlook. He examined the history of a country within a large geographic area, not a city, strove to obtain a broad, interpretive framework, and showed some of the connecting features in various aspects of Italian experience.

Though such historians addressed modern problems, methodological shortcomings still impeded the development of a modern historical consciousness. Scholars had not yet developed appropriate methodological techniques for elucidating the meaning of historical artifacts within the context of their own times, places, and cultures. To be sure, antiquarians and humanists during the Renaissance collected and preserved Greek and Roman documents and by so doing inadvertently created repositories of immense value for future historians. But the more immediate interest centered on venerating the memory and emulating the style than in subjecting the ancients to analysis and interpretation. Philology, the science of verifying and authenticating old manuscripts, preoccupied them more than history. In this endeavor, critical techniques, indeed, made possible the identification of forgeries, alterations, and mistakes in transcription. A famous example involved a political fraud. According to a widely accepted myth in the late Middle Ages, Constantine the Great, the first Christian emperor of Rome, as an act of humility upon his conversion, had bestowed the right to wield religious authority over the Roman Empire in the West upon the bishop of Rome, in this case Pope Sylvester II. The popes in later years had based in part their claim to power upon the legend. In 1440, Lorenzo Valla proved the story false in a brilliant piece of work entitled *Discourse on the Forgery of the Alleged Donation of Constantine.* Using the methodologies of textual criticism, Valla demonstrated that the presumed substantiating document abounded with errors, inconsistencies, and anachronisms, showing that no one could have written the *Donation* in the fourth century. The fake had to come from a later time.

The Protestant Reformation and the ensuing religious and political upheavals in the sixteenth and seventeenth centuries in Europe intensified the interest in history among the warring factions but made it the agent of partisan and polemical debates. Antagonists on all sides invoked the sanction of the past to give authority to their positions in the present. The issues centered on the role of the Church in interpreting the word of God to believers, the character of the Bible as a historical text, and the circumstances surrounding the development of Christianity as an institutional religion.

Protestants attacked their adversaries by insisting that papal control had corrupted the purity of belief and practice inherited from the early Church. In Germany, Martin Luther and his ally, Philipp Melanchthon, used history to challenge Roman Catholic traditions; and somewhat later, a group of Lutheran scholars under Matthias Flacius Illyricus published a study of thirteen volumes entitled *Magdeburg Centuries*, an effort to document Lutheran assertions that the Roman hierarchy had distorted and perverted the teachings of Christ and the apostles. In a similar vein, Robert Barnes, an Englishman, wrote *Lives of Roman Pontiffs*, in which he ascribed guilt and vilification to all. Roman Catholics fought back, using history in defense of their beliefs and practices. In 1571, a papal commission chose Caesar Boronius to write a rebuttal to the *Magdeburg Centuries*. His *Ecclesiastical Annals* argued that postapostolic changes had legitimacy as clarifications and interpretations of Christ's teachings and, moreover, had taken place under the guidance of the Holy Spirit.

Such controversies divided all of Christendom and resulted in the establishment of some of the first professorships in history at European universities. The Lutherans, for example, established one at Heidelberg, and the Calvinists another at Leyden. Though over the long term, such turns of events would advance the cause of professional history, the more immediate impact contributed to the breakdown of uniformity in Christian interpretations of history and perhaps also to a growing gap between sacred and secular understanding. As a consequence of the Reformation, the consensus in support of St. Augustine's approach to universal history disintegrated, and no new one emerged immediately to replace it.

Meanwhile, the great scientific revolutions of the seventeenth century produced another danger by bringing into question the very possibility of obtaining reliable accounts of the past. Following the discoveries of Newton, Kepler, and Galileo within the orders of nature, a scientific world view took hold of European intellectuals, many of whom doubted the feasibility of obtaining verifiable knowledge in imprecise fields of investigation such as history. For philosophers schooled in the wonders of scientific discovery such as Descartes, Spinoza, and Leibnitz, mathematical formulations held the key to exactitude and certainty. As B. A. Haddock observed, "Mathematics had removed the aura of mystery from nature, had freed enquiry from the obfuscation of medieval obscurantism, and it was thought that the same tool would be similarly successful in the study of man."

According to such champions, if historians could not express knowledge in approximate conformance with the mathematical ideal, then they had either embarked upon a "harmless but irrelevant enjoyment of confused perceptions" or, much worse, constituted "a dangerous error in the path of truth." As Haddock noted, for the philosophers, "These mechanical methods were essentially timeless. They used a model of reason, eternally valid, based on their own mathematical procedures, which destroyed the validity of all other modes of experience."[1] By so doing, they set before historians one of the great philosophical issues of the last four centuries, the degree to which the natural sciences can and should determine the forms of knowledge in the study of human affairs.

In spite of the religious and philosophical hue and cry, more pragmatically minded historians produced significant and enduring works. In 1566, Jean Bodin, a French teacher of jurisprudence, published *Methods for the Easy Comprehension of History*, in which he called for an expansion of legal studies in order to escape from the prevailing provincialism and to create a new kind of universal history, resulting supposedly in comparisons of the laws and customs of all nations. Though Bodin's complex categories entailed distinctions among divine, natural, and human history, he organized his material clearly and precisely according to chronological and geographic criteria, and he insisted also upon the preeminence of primary over secondary sources. A little over a century later in 1681, another Frenchman, Jean Mabillon, published two volumes entitled *On Diplomatics*. Pronounced a masterpiece in the twentieth century by no less an authority than the great French medievalist Marc Bloch, this study gave instruction in methodological rigor by providing a treatise on the methods and techniques of deciphering ancient charters and manuscripts and determining their authenticity, a science then referred to as "diplomatics." The capacity to detect forged documents, of course, had great practical significance. Mabillon insisted that knowledge, care, and an eye for the inconsistent and out-of-place could result in such determinations through a process of comparison.

But such tenacious scrutiny of sources was uncharacteristic. In the same year, 1681, Bishop Jacques Bénigne Bossuet wrote a *Discourse on Universal History* as a means of teaching morality to the Dauphin, the male heir of King Louis XIV. Unchanged much thematically from St. Augustine, the *Discourse* affirmed tradition and authority by employing a providential view and depicting the Roman Catholic Church as the chosen agent of God's will. A kind of throwback, this work really anteceded the Reformation, the scientific revolution, and the methodologies of diplomatics.

The era of the Enlightenment produced a torrent of historical writing and affected historical thinking significantly. In the eighteenth century, European philosophers proclaimed the advent of a new age of advancement for humankind. The attainment of enlightenment required an end of ignorance and superstition and dictated that henceforth the faculty of reason should govern the conduct of human behavior. As a corollary, the movement implied a rebellion against the authority of traditional religion. In a famous and important book, *The Idea of History*, the British philosopher and historian Robin G. Collingwood characterized the Enlightenment as an attempt "to secularize every department of human life and thought." He saw it as "a revolt not only against the power of institutional religion but against religion itself." As Collingwood explained, Voltaire, a leading figure, regarded himself as the spearhead in a crusade against Christianity and other backward and barbarous influences. His slogan, "Ecrasez l'infâme," called upon his contemporaries to "crush the wicked thing."[2]

Similar tendencies also appeared in the English-speaking world. In 1586, William Camden published *Brittania*, a study of pre-Roman antiquity. It anticipated some of the functions of later scholarly productions by debunking as false the heroic tales of King Arthur's time and also by espousing strong nationalism. According to one commentator, Camden "wanted to be both patriotic and scientific . . . an enormous assignment, but he succeeded magnificently."[3] About a generation later, another Englishman, albeit a transplanted one, engaged in an impressive exercise in writing contemporary history. William Bradford began *Of Plymouth Plantation* in 1630 and continued it for two decades. This spare, simple narrative provided the most complete story of the Plymouth colony on Cape Cod during the early years. Bradford served as governor for a long time. His stark tale of anguish and adversity recorded the trials endured by the settlers and incorporated a message regarded by all Protestant Pilgrims as true. The hand of God directed human history in inscrutable ways. Another example of writing contemporary history emerged from the great religious and political struggles between the Puritans and the Cavaliers in the middle of the seventeenth century. This subject over the years has preoccupied many students of English history. An early account by Edward Hyde, the first Earl of Clarendon, entitled *History of the Rebellion and Civil Wars in England*, presented a passionate defense on behalf of the Church of England and the king against the assaults of tyranny, regicide, and irreligion.

Such issues produced deep divisions. The foremost historians—Voltaire, David Hume, and Edward Gibbon—disparaged the role of religion and depicted it as an impediment to human progress. For Voltaire, the priestly classes ranked among the great deceivers throughout the ages, the purveyors of bigotry, intolerance, and oppression. One mark of advancement called for the liberation of people from such benighted captivity. For Enlightenment historians, the course of history compelled a march toward rationality and emancipation.

Attempts to move beyond the customary categories also typified the writing of history during the Enlightenment. The secular confines ordinarily comprised political and military affairs and centered on the activities of western Europeans. Voltaire most notably wanted an extension of the frontiers to encompass social, economic, and cultural dimensions and also the inclusion of non-Europeans. Indeed, he consciously composed some of the first social and cultural history and developed a kind of fixation on the Chinese.

In spite of such laudable impulses, much of the history during the Enlightenment suffered from a fundamental flaw. It consisted of an incapacity to comprehend the behavior of historical actors on their own terms. The historians lacked a truly historical sense of development in the past. Locked into the precepts of their own time, they regarded their own values and aspirations as universal and absolute, the best toward which humankind could strive. Consequently, they tended to regard deviations in other times and other places either as aberrance or folly.

Collingwood made the point pungently, arguing that "the historical outlook of the Enlightenment was not genuinely historical." Rather, he described it as "polemical and antihistorical." Paradoxically, the past produced repugnance among the historians. They interpreted it as something from which to escape, in Collingwood's words, "sheer error, . . . a thing devoid of all positive value whatever." As a result, they had difficulty comprehending the past as the participants experienced it, and when gauged against the standards of the present, it failed to measure up. A related problem stemmed from a reluctance to carry out much archival research. Instead, the historians preferred to draw on existing works. As Collingwood remarked, "they were not sufficiently interested in history for its own sake to persevere in the task of reconstructing the history of obscure and remote periods."[4]

Voltaire, the very embodiment of the Enlightenment spirit, provided the personification. The composer of an immense number of works in many different literary genres, he also figured prominently in the development of historical writing. First and foremost, he insisted that history should have a practical purpose. Pursued reflectively and philosophically (*en philosophe*), the study of the past should contribute to the cause of enlightenment by freeing readers from fallacy and misconception. Voltaire affirmed the point bluntly, claiming that "All ages resemble one another in respect of the criminal folly."[5] For the same reason, presumably, all merit historical study. Only through the act of facing up could release come about. Voltaire intended further to write discriminating history, concentrating especially on "that which deserves the attention of all time, which paints the spirit and the customs of men, which may serve for instruction and to counsel the love of virtue, of the arts and of the fatherland."

Voltaire's principal historical works consisted of *Philoso-*

phy of History, later incorporated as an introduction into *Essay on the Customs and the Spirit of Nations*, and his masterpiece, *The Age of Louis XIV*. In the first essay, employing in title a term he coined, Voltaire intended to illuminate the mind of his mistress, Madame du Châtelet. By depicting philosophically the human advance out of barbarism, he showed the consistently malignant influence of priests among the Egyptians, Hebrews, and Christians. In effect, he turned medieval historiography on its head. But Voltaire put little stock in basing his claims on evidence. For him, reason and credibility, more than original documents, should provide the test of truth. For example, he questioned the story of Noah's ark by asking "how many persons were in the ark to feed all the animals for ten whole months, and during the following year in which no food would be produced?" The second volume depicted the late seventeenth century, the age of Louis XIV, as one of four ages of "great attainment" in the history of humanity and hence an object of special attention. The other three took place in Greece under Pericles, in Rome under Caesar and Augustus, and in Constantinople after the fall of the city to Mahomet II. His portrait of the many achievements in Europe during the reign of Louis XIV amounted to the invention of cultural history.

Though less a polemicist than Voltaire, David Hume, a Scotsman, shared the secular bent. Contemporary critics denounced his philosophical works as the products of skepticism and atheism. In contrast, his historical works won him fame and fortune. His six-volume *History of England* was the best and most popular English history in the eighteenth century. Moving backward in time, Hume wrote first about the Stuart kings, then the Tudors, and finally the events since the invasion of Julius Caesar. Critical and iconoclastic in approach, Hume wrecked romantic legends and patriotic myths. As a rigorous empiricist, he hesitated always to claim more than the observation of phenomena could warrant or to think that patterns of the past necessarily recur in the future. Hume consequently found in history no particular sanction for any specific moral or legislative programs. Indeed, he mistrusted metaphysical claims and religious dogmatism as the breeders of fanaticism and instability. For him, the revolution against King Charles I amounted to a case in point. Political equipoise required an ability to adapt institutions to the dictates of experience.

Perhaps the greatest of the Enlightenment histories, *The Decline and Fall of the Roman Empire* by Edward Gibbon, covered a period of 1,300 years from the first century until the Renaissance and ran to some million and a quarter words. Gibbon, an Englishman, spent twenty years of his life on the project but invested little effort in searching out new documents. Instead, he drew heavily on the works of seventeenth-century scholars and antiquarians. This colossal effort amounted to a synthesis of the existing research on the ancient and medieval worlds. It also transposed St. Augustine's thesis in *The City of God*. Rather than absolve the Christians of responsibility in bringing about the terrible events leading to Rome's fall, Gibbon held them primarily accountable to bear the burden. To simplify his argument exceedingly, Gibbon maintained that the obsessions of a hard-pressed people with life after death resulted in the neglect of imperial interests and the need to maintain proper defenses. Other-worldly concerns corroded the Romans and brought about the collapse. In a kind of summary statement, Gibbon set forth the essentials of his view:

> The theologian may indulge the pleasing task of describing Religion as she descended from Heaven, arrayed in her native purity. A more melancholy duty is imposed on the historian. He must discover the inevitable mixture of error and corruption which she contracts in a long residence upon earth, among a weak and degenerate race of beings.[6]

This kind of condescension toward the past raised important methodological questions. Critics such as Robin G. Collingwood in the twentieth century attacked Enlightenment historians on the grounds that their insensitivity in effect violated the integrity of history. More specifically, they failed to empathize properly with the historical actors and to comprehend their behavior accurately on their own terms. Rather, Enlightenment scholars indulged in exposés, reviling the past to obliterate and overcome it. Consequently, Collingwood denounced their writing as an enterprise gone fundamentally wrong. They had failed to carry out the historian's primary task, that is, to elucidate the past, not merely to condemn it. For Collingwood, the magnitude of irresponsibility took on even greater proportions because a methodological means for achieving the goal already existed. Giambattista Vico, an obscure Italian theorist from Naples, had already shown the way.

Vico published his brilliant and innovative book in 1725. Hailed in the twentieth century as the product of genius, truly an anticipation of modern social science, *The New Science* had much less impact on the eighteenth century. In discouragement, Vico himself lamented of his work, "it has gone forth into a desert." No one initially paid much attention, even though Vico had provided the methodological means for resolving the dilemma of the Enlightenment historians. To overcome the confines of one's own time and to treat the past with a proper degree of historicity, he demonstrated that scholars had an obligation to reconstruct the mental universe of people in previous times in order to account for their actions. Voltaire could have learned a great deal from Vico.

Vico presented *The New Science* as a critical response to the overblown and exaggerated claims in natural science. Unlike René Descartes, who had disenfranchised history as a branch of knowledge because of the notorious unreliability of traditional accounts, a manifestation of fantasy and make-believe, Vico established an epistemology by which to set forth verifiable, true claims. While refraining from any denial of the validity of mathematics, he insisted upon the feasibility of other forms of knowing. To rebut Descartes' claim that the natural order constituted the most promising field for truthful inquiry, he insisted upon an intriguing idea, really, a kind of first principle. To understand anything fully, he argued, the observer must have made the thing under observation. Since God had created nature, only he could grasp the totality of it. In contrast, since human beings had made history, they possessed full capacity for arriving at a correct understanding, providing, of course, that they employed the right methods.

Vico conceived of history as a dynamic process of change, characterized by a movement through three stages. Though admittedly Eurocentric in design, the scheme at least rested on a recognition that different people in different places and different times actually saw and experienced the world differently. As a devout Christian, Vico adhered to the essentials of the Old Testament rendition until the aftermath of the Great Flood but then put his own stamp on the course of events. Noah's descendants, after some two hundred years of chaos, embarked upon a journey through the age of gods, the age of heroes, and the age of men. Vico believed that human nature underwent changes during the progression and, moreover, that identifiable links connected beliefs and institutions into a whole. During the first, an age of sensation, ferocity and cruelty typified human nature, and governments took on theocratic forms, thereby providing appropriate subordination to prevailing conceptions of deity, regarded as the cause of all things. During the second, an age of imagination, nobility, and pride constituted human nature, and the government consisted of a warrior aristocracy. During this time, language, law, and culture developed, and mythic or poetic modes of consciousness became dominant. Finally, during the third, an age of reason, human nature emerged as reasonable and benign; governmental institutions evolved into egalitarian democracy, and a sense of reflective rationality governed human behavior. But Vico would not bask in pure optimism. He anticipated the possibility of returns to barbarism by imagining history to move in a spiral with cycles as part of the process.

The means of proper study emanated from philosophy, that is, reasoning from axioms, definitions, and postulates, and from philology, the empirical study of the languages, history, and literature of people. Oddly, considering Vico's animus against Descartes, *The New Science* assumed the format of a geometry text, setting forth specific rules bv which to obtain the truth. Among other things, they warned of prejudices conducive to error. Scholars should beware of exaggerating the grandeur of the past. They should recognize that all nations favor their own history at the expense of others. They should not assume that people in the past resembled them or that the ancients knew more about themselves than current scholars. Vico proposed to employ the study of linguistics, mythology, and law as primary keys to unlock the meaning. Historians could never comprehend their predecessors unless they learned to see them as the people viewed themselves.

The roots of words, tales, myths, legends, and legal systems all conveyed important clues. For Voltaire, a story about supernatural beings acting upon human residents of the world signified ignorance and superstition. For Vico, it provided a means of investigating those peoples' conceptions of themselves and their position in the universe. Vico produced an early version of some of the precepts of "historicism" in the nineteenth century. While Enlightenment thinkers liberated history from the constrictions of theology, Vico and his followers discovered methods by which actually to recover the past. In the nineteenth century, their disciples obtained a modern consciousness.

History in the nineteenth century became diffuse, multifaceted, and ever more rigorous methodologically. As a consequence, the study of history obtained respectability in the great European universities of the day and achieved the status of an academic discipline. Professionally trained historians ultimately dominated the field, but gifted amateurs and philosophers with an interest in history also contributed to three distinctive kinds of historical writing. One strand grew out of the struggles of the French Revolution and the Napoleonic Wars and emphasized a romantic and nationalistic approach. A second, most characteristic of Germany, perceived the flow of history as a subject for abstract, philosophical thinking. A third, a forerunner of today's university-based, professional history, aspired to find out what actually happened through careful investigations based upon archival research. The ensuing process of intellectual enrichment resulted in an era of unprecedented attainment and overcame the most incapacitating shortcomings of Enlightenment history. Rather than manifest disdain for the past, the historians of the nineteenth century reveled in its diversity.

Romantic and nationalistic historiography predominated in France after the Bourbon Restoration in 1815. Jules Michelet was a prime exemplar. As a self-conscious disciple of Vico, Michelet aspired to write the history of the French people. While using unconventional sources such as folklore, songs, poetry, and architecture, he proposed to develop portraits of everyday life. He identified with the common people. His six-volume *History of France* depicted and eulogized peasant life in the medieval period in vivid detail. It also exulted in the emergence of a sense of nationhood. The patriotism of another romantic nationalist, François Guizot, won fame for the historian and also the position of first minister under King Louis Philippe. History and politics served nicely together in Guizot's case.

In England, Thomas Babington Macaulay also favored romance. As a prose stylist he had few equals. High-blown and self-consciously magisterial in the approved fashion of the nineteenth century, Macaulay's *History of England* obtained a wide audience and established history as a branch of great literature. It also articulated an uncompromising version of the Whig interpretation, treating the nation's story as synonymous with the emergence of liberty. As he wrote, "The history of England is emphatically the history of progress. It is the history of a constant movement of a great society." In sonorous and balanced cadences, he argued the case, applauding the triumph of the Glorious Revolution of 1688 over the Stuart despotism, concluding that "the history of our country . . . is eminently the history of physical, of moral, and of intellectual improvement."[7]

In the United States, the works of George Bancroft, Francis Parkman, and Henry Adams also set forth the themes of romantic nationalism. These great amateurs employed soaring prose for dramatic effects. Bancroft, like Guizot, combined careers in scholarship and public service. His writings told the story of national development, emphasizing the advance of liberty as recounted from the viewpoint of a Jeffersonian Democrat. Parkman's elegant prose reconstructed in a series of books the conflict between France and Great Britain in North America during colonial times. *The Oregon Trail*, another sort of adventure, created a kind of nature poem. Henry Adams's nine-volume *History of the United States during the Jefferson and Madison Administrations* aspired to the status of science. As the grandson of one president and the great-grandson of another, the cosmopolitan and well-connected Adams wanted to tell the nation's story and also to uncover the laws governing invariable relationships in human affairs. Although later in his life he judged this part of the effort a failure, the work still ranks as a foremost scholarly attainment in the nineteenth century.

Abstract, philosophical history emanated primarily from Germany. It had deep roots in the eighteenth century. Johann

Gottfried Herder and Immanuel Kant both wrote philosophical treatises on history. Herder's *Ideas toward a Philosophy of History of Man* described human life as closely related to nature and portrayed history as an evolutionary process. Each stage in the course of human development depended upon everything that had gone before. Otherwise, the latter stages could not occur at all. What existed in the present and would come about in the future necessarily presumed what had been. This claim demolished the Enlightenment's habit of disparaging the past and compelled historians to appreciate the integrity of all eras. Herder particularly exulted in the varieties of human experience and extrapolated an important observation from them. According to Collingwood, he was "the first thinker to recognize in a systematic way that there are differences between different kinds of men, and that human nature is not uniform but diversified."[8] This view anticipated Collingwood's own conviction that no people have a fixed nature. Instead, they have their history and become what historical experience makes of them. Collingwood held that true historical thinking can never exist without this rudimentary recognition.

Immanuel Kant, the elder of the two men, reacted negatively to such claims and wrote a critical response entitled "An Idea for a Universal History from a Cosmopolitan Point of View." Though formerly Herder's teacher, Kant thought the work of his student wrongheaded and intended to provide a correction. Though concerned only tangentially with the study of history, he too called for the philosophical treatment of the subject. Harkening back toward Enlightenment traditions, he rejected Herder's emphasis on human dissimilarities and proposed instead to regard the unfolding of history as "the realization of a hidden plan of nature" by which "all the capacities implanted by her in mankind can be fully developed." Less impressed with diversity, Kant dwelled upon uniformity. For him, history amounted to a process by which humankind became rational and hence fulfilled its fundamental nature. Consequently, the main task of the philosophical historians required that they comprehend the means and mechanisms for attaining the aim.

Georg Wilhelm Friedrich Hegel addressed the issue in an ornate body of writing intended to explain the whole of the experienced world. Defining nature and history as the manifestations of divine will in space and time, Hegel developed a complicated and inclusive philosophical system to make the whole intelligible. Its very complexity defied summary and often put off practicing historians who doubted the usefulness and verifiability of Hegel's magnificent abstractions. His *Philosophy of History*, a series of lectures compiled and published after his death, pursued no less an ambition than to make the entirety of the human past comprehensible.

To begin a brief characterization, Hegel espoused profound religious commitments. In repudiation of Enlightenment thinking, he put God back into history through his interpretation of it as a logical and orderly process, resulting in the manifestation of divine will in time. As a consequence, the attributes of reason and freedom, identified as characteristics of divine spirit, would gradually encompass the world. To say it in a different way, Hegel wanted to provide a kind of theodicy in which he would explain the ways of God toward his creation. For Hegel, the source of experienced reality first existed as an abstract ideal, later made actual in the world through the unfolding of divine will in time.

In Hegel's scheme, the mechanism of change bringing about the realization of divine will proceeded dialectically, and human minds could arrive at understanding by reasoning similarly. In his world of pure thought, all things gave rise to their opposites. Light had no meaning apart from dark. The same held true for high and low, sharp and dull, near and far. In like fashion an idea, or a thesis, resulted in a counter proposition, or antithesis, and the ensuing debate produced a synthesis, which in turn became a thesis and set the whole thing in motion again. For Hegel, the dialectic not only characterized how thought progressed toward understanding but also how change actually took place in the material world. Confusingly, he added that the steps could take place out of logical sequence.

The march of history, nevertheless, produced changing levels of consciousness among human beings. In effect, their natures underwent alterations as a result of new forms of experience and awareness. To put it simplistically, they became more rational and more free. As Hegel explained, no people could attain the ends of history, that is, freedom and rationality, until they knew that they possessed the capacity to exercise such faculties. The consciousness dawned slowly. As Hegel observed, "the East knew and to the present days knows only that *One* is free [i.e., the oriental despot]; the Greek and Roman world, that *some* are free [i.e., the aristocratic elite]; the German World knows that *All* are free [under the legal structure of the monarchy]."[9]

Though Hegel believed that he had explained the forms and logic of historical development with proper regard for the facts, not many practicing historians ever adopted the Hegelian system or tried to work within its constraints. Nevertheless, the effects significantly influenced historical thinking in the nineteenth century. Most obviously, though Hegel shared with Kant an inclination to emphasize emergent rationality among humans, in many ways an equivalent of freedom, he also sided with Herder in his comprehension of human nature as malleable and bending, never fixed or final. It existed in a condition of becoming. Accordingly, different humans in different times and places had quite different natures. Hegel also favored Herder's holistic conception of the human past in which each phase maintained integrity as a prerequisite to whatever followed. For Hegel, the patronizing attitudes of the Enlightenment could only distort the past and prevent true understanding. As a means to obtain it, he insisted that historians must study bygone ages and epochs on their own terms.

These two points became central assertions of a school of historical thinking in nineteenth-century Germany. Termed "historicism" in English, an awkward translation of the German word *historismus*, this approach affirmed the need for particular methodological means to fathom the meaning of the past. Most fundamentally, it pointed to the diversity of human experience and claimed that as a consequence different peoples quite literally viewed the world differently. To comprehend their world, scholars had to enter into their mental universes empathetically and reconstruct their pictures of reality. Only then could scholars credibly explain their forms of behavior. Knowledge of how they thought would facilitate an understanding of how they acted.

German historicists operated out of universities and spent their time in archives working with documents. By and large, they accepted Hegel's methodological advice but ignored or rejected his larger metaphysical system. It probably struck them as an a priori invention, neither proven nor provable. Rather than embark upon grandiose ventures, most of them embraced the more modest aim of writing history as it actually happened, or, to use the language of one of the foremost practitioners, *wie es eigentlich gesewen.* Leopold von Ranke wanted to describe historical events "as they really were." More than anyone, he transformed history into a modern academic discipline, university-based, archive-bound, and professional insofar as the leading proponents underwent extensive postgraduate training. The methods and techniques emanating from Germany spread elsewhere upon the continent, to Great Britain, and to the United States. They called for extensive research in primary sources to discover the truth, for detached and unbiased judgments, and for a determination by historians to see and to experience the world as it seemed to the historical actors.

Ranke's life work staggers the imagination. He lived ninety years and produced sixty volumes of published works. As the inventor of modern history, he chose the emergence of the European state system after the Reformation as his great theme. Working out of the University of Berlin for fifty years, his seminar method of instruction and his prolific scholarship made him world famous. George Bancroft of the United States on one occasion called him "the greatest living historian." Ranke stated his purpose in the "Preface" to his *Histories of the Latin and Germanic Nations from 1494–1514.* "History has been assigned the office of judging the past, of instructing our times for the benefit of future years. This essay does not aspire to such high offices; it only wants to show how it had really been—*wie es eigentlich gewesen.*"[10]

Ranke aspired to achieve balance and objectivity. Though on occasion his personal commitments to monarchy, Protestantism, and the Prussian nation state perhaps colored his judgments, they never destroyed them. Indeed, his works retain believability until the present day because of his comprehensive research and his scholarly disengagement. For example, his masterpiece, the *History of the Popes,* aroused criticism among contemporaries who thought it too bland and impartial. He also displayed great breadth. At age eighty-three, he set out to compose a universal history and managed to get through the fifteenth century before his death in 1885. For such prodigious effort and achievement, Ranke's legacy persisted into the twentieth century and profoundly shaped the writing of professional history.

Another strain of German thinking also retains an impact in the present day. Also acting out of repudiation, the young Karl Marx encountered Hegel's idealism while pursuing advanced studies in philosophy at the University of Berlin. Though enamored of the dialectical construct as a means of describing change, Marx could not accept the abstract and ethereal character of the Hegelian system. For Marx, reality resided in the physical world, and its imprints shaped the minds of humans, not the other way around as Hegel believed. Marx's work throughout his life aimed at rebutting Hegel's claims and establishing a new mode of philosophical understanding founded upon dialectical materialism.

Marx focused attention on how people earned their livings. For him, work constituted a central facet of life and prime determinant of all the others. The technologies of work also fascinated him. He believed that he could explain a great deal by examining economic relationships and finding out who owned the tools and agencies of production and who provided the labor in the fields and the factories. In Marx's scheme of things, an unavoidable and irremedial conflict of interest set the two apart, because the former necessarily exploited the latter by appropriating a disproportionate share of the wealth produced through workers' endeavors. Conceptions of class figured prominently in his analysis, and he defined them mainly in economic terms, insisting that the whole of human history consisted of a struggle between those who possessed more than enough to satisfy their needs and those who as a consequence never ceased to want.

Much of Marx's writings sought to demonstrate the validity of such claims through investigations into the workings of European capitalism, but the scholar also functioned as a prophet, promising assurances of deliverance. For Marx, history consisted of a sequence of tumultuous upheavals in which the established systems of production underwent dramatic alteration and consequently unworked all of the existing economic, social, and political relationships. Change, the enduring constant, proceeded dialectically and marked the measure of advancement, bringing human beings ever closer to a more perfect future in which all people would share more equitably in the rewards of work. His visions drew upon important legacies in the philosophy of history, running backward in time to the Jews and the early Christians, but in this rendition, the very course of history rather than transcendence over it would bring about salvation. Affirming a linear and progressive view of historical development, Marx put philosophy of history in service to revolution. As he remarked as a young man, "the philosophers have only *interpreted* the world, in various ways; the point, however, is to *change* it."[11]

The Marxist injunction created immense divisions in the modern world over the proper understanding of politics, economics, and the role of history. While Marx and his followers issued the call to revolutionary action and sought legitimation in historical imperatives, more conventional and traditionalistic scholars held back from such leaps, refusing to employ their mode of investigation as justification for any vision of the future. Characteristically trained in the graduate schools in the importance of archival research and methodological and conceptual rigor, they hesitated to extrapolate their findings into the future, preferring to restrict their professional interests to presumably knowable subjects. To an extent out of caution and relief, they left the predictive function, that is, ruminations about the prospects still to come, to philosophers of history who addressed it in imaginative and challenging ways.

RECOMMENDED READINGS

Historians at Work, Vol. 2: *From Valla to Gibbon*, and Vol. 3: *Niebuhr to Maitland*, edited by Peter Gay and Victor G. Wexler (New York: Harper & Row, Publishers, 1972, 1975) make available samples of the writings of the great historians. Herbert Butterfield, *The Origins of History* (New York: Basic Books, Inc., Publishers, 1981) traces the process of secularization. A general commentary is provided in Paul K. Conkin and Roland N. Stromberg, *The Heritage and Challenge of History* (New York: Dodd, Mead & Company, 1971). R.G. Collingwood, *The Idea of History* (New York: Oxford University Press, 1956) is indispensable. *History and Historians in the Nineteenth Century* by G.P. Gooch (Boston: Beacon Press, 1959, first published in 1913); and *The Theory and Practice of History* by Leopold von Ranke, edited by Georg C. Iggers and Konrad von Moltke (New York: The Bobbs-Merrill Co., Inc., 1973) vividly depict developments during "the golden age" and make them accessible to English-language readers. Leonard Krieger, *Ranke, The Meaning of History* (University of Chicago Press, 1977) is a sophisticated intellectual biography which employs the techniques of psychohistory. Finally, B.A. Haddock, *An Introduction to Historical Thought* (London: Edward Arnold, Publishers, Ltd., 1980) provides a discussion of modern historical thought, particularly over the question whether the models of natural science provide an apt basis for the study of history.

4

PHILOSOPHY OF HISTORY: Speculative Approaches

The term "philosophy of history" has acquired several connotations during the past three centuries. When Voltaire coined the term by insisting upon history done *"en philosophe,"* he meant the construction of more meaningful narratives through the exercise of thoughtful reflection. Later the phrase took on different definitions. One encouraged speculation over the course and aims of history. This approach, the subject of this chapter, sought to obtain profound levels of truth by discerning patterns in the past and connecting them with expectations for the future. It required explanations that would aid in foretelling the goals and purposes of history. A second version addressed in the next chapter, the analytical philosophy of history, has focused on problems in methodology, particularly in the areas of logic and epistemology. This approach has set forth rigorous standards and sometimes has scrutinized the degree to which the actual practices of historians have measured up. In combination, the emphasis on imaginative inquiry controlled always by reason and evidence has contributed to intellectual richness in the discipline.

Human beings have engaged in ruminations about the directions of history since ancient times. Among other things, they have sought to mute their sense of vulnerability in facing the unknown by seeking to determine recurring tendencies in the past. Over the years, at least three basic schema have characterized such endeavors, namely the cyclical, the providential, and the progressive. To be sure, variations on each have occurred, and imprecisions of usage have permitted a degree of overlap and intermixture. But in spite of such confusions and indistinctions, the first view in fundamental form has depicted history as a motion in circles, repeating endlessly over and over again. The second and the third, in contrast, have conceived of a process moving through time in a linear fashion from a clear beginning to a middle to an end. Though each of these two versions has presumed advancement, the motive force has taken different forms, perceived in the first instance as the product of divine guidance and in the second as the result of metaphysical or natural forces impelling progress.

Manifestations of the cyclical idea cropped up among many ancient peoples and in some Eastern cultures. Egyptians, Sumerians, Babylonians, and Greeks by and large derived the premise from the orderly fluctuations in nature and attached little real significance to the passage of events. In a sense, they refused history. As noted by a historian of religions, Mircea Eliade, in his book *Cosmos and History: The Myth of the Eternal Return*, the inhabitants of traditional or "archaic" societies lived and operated in a kind of timeless present, a world very different from that apprehended by Jews and Christians. Though cyclical interpretations persisted even into the twentieth century, for example a dazzling and perplexing work entitled *The Decline of the West* by Oswald Spengler, a German philosopher and historian, the more characteristic form of thinking in Western civilization has emphasized linear conceptions of the past.

A belief in providential design, affirmed so exhaustively by St. Augustine in the fifth century, appeared incessantly in the historical writing of the Middle Ages. This context gave coherence to medieval chronicles by providing the means to fit contemporary records into the flow of universal history and also to express the faith in the reality of God's presence and purpose. Sometimes unique formulations resulted in dramatic twists. Late in the twelfth century, Joachim de Flora divided history into three epochs, each corresponding to a figure in the Holy Trinity. The Age of the Father encompassed the pre-Christian era, the Age of the Son, the Christian era, and the Age of the Holy Spirit, the anticipation of the future. According to Joachim, each period would endure for forty generations, and the passage from one to the next would signify advancements toward conditions of greater wisdom, love, spirituality, and happiness. The prophecy ironically took on revolutionary political implications after Joachim's death in 1202 by sparking interest among poor and humble people who aspired to a better life and resulting in an upsurge of religious millenarianism.

A tripartite division of the past also appeared in Giambattista Vico's *The New Science*. A devout Roman Catholic, Vico tried to combine his belief in divine providence with a scholarly insistence upon the need for empirical investigations. Demonstrating the facts in the particular presumably would confirm the veracity of the larger whole. In his scheme of things, an amalgam of elements, history advanced progressively through stages but also fluctuated in cycles, resulting in a kind of spiral effect. Vico perceived a three-tiered progression from "the age of the gods" to "the age of heroes" to "the age of men," and in each he detected holistic relationships among institutions, practices, and beliefs. Together they made up a uniformity. In his words, in the first stage, "the gentiles believed they lived under divine governments and everything was commanded them by auspices and oracles." In the second, "heroes ... reigned everywhere in aristocratic commonwealths, on account of a superiority of nature which they held themselves to have," and in the third, "all men recognized themselves as equal in human nature and therefore there were established first the popular commonwealths and then the monarchies, both of which are forms of human government."[1]

Vico's extrapolation from these representations held that special forms of government, theocratic, aristocratic, republican, or monarchic, corresponded to each change, and moreover, that particular kinds of language and jurisprudence necessarily resulted. For Vico, the various cultural parts interlocked at each state, forming a coherent and uniform whole. In a kind of summary statement, he intimated that the shifts further reflected developments in the makeup of human beings, affirming that "The nature of peoples is first crude, then severe, then benign, then delicate, finally dissolute." The cycles necessarily would run the course, each giving cultural expression to successive changes, and then recurring on a higher plane, resulting in a progressive movement under God's supervision.

In the nineteenth century, Georg Wilhelm Friedrich Hegel endorsed some similar propositions. Indeed, he calculated the whole of his philosophical system to demonstrate the creative immanence of divine power in space and time. Nature and history for him served up the testimony. In what may have amounted to the most ambitious philosophical treatment of history ever, Hegel set forth his views, defining a dynamic and rational process aimed at the attainment of reason and freedom. For Hegel, the course of history moved by comprehensible means toward the stated goal in a dialectical fashion. That is, the mechanisms of historical change required that ideas, or mind, generate reality by giving rise to opposites so that a constant tug-of-war went on between that which actually existed and that which might come about.

In somewhat different words, according to Hegel, history consisted of the means by which God achieved his purposes in the world and human beings arrived at new and higher levels of consciousness, both of their surroundings and of their own capabilities. The course of development moved generally from East to West from the Orient through the intervening regions to western Europe and took on institutional forms in the great religions and the apparatus of the state. Hegel espoused a somewhat crankish Eurocentric view, excluding Africa, the New World, and the Slavic domains from history on the grounds that the inhabitants had contributed little to the progress of humankind. In contrast, European regions, Greece, Rome, and the Germanies, had participated conspicuously. Hegel assigned special importance to World Heroes, his term for great men, who through the pursuit of their own interests actually served larger purposes. In quest of personal empire, Napoleon promoted European progress by sweeping away feudal remnants. Hegel attributed the unintended outcome to "the cunning of reason," his phrase for the capacity of divine power to manipulate historical actors in service to its own purposes.

The power of Hegel's intellect resulted in formidable attainments. The system defined a theodicy, that is, a justification of God's way toward humankind, and also, in Hegel's own view, conformed with empirical proofs. Yet, more practical-minded historians regarded such constructs as the result not of hard facts but of unproven assumptions and a priori reasoning. The overwhelming inclusiveness of his philosophy of history, that is, his fierce determination to account for all things, would have placed unbearable strains on more orthodox historians seeking to work within the constraints. Very few scholars ever tried to compose their narratives within the confines of Hegel's mental universe. Nevertheless, his contribution to historical thinking figured prominently, particularly in the emergent German historicist school, by depicting the flow of history as consisting of successive changes in the level of consciousness among human beings.

A shift toward secular characterizations of the past in progressive, linear terms achieved particular prominence during the Age of the Enlightenment in the eighteenth century. Typically these approaches set forth a preferred and favored vision of the future by which to bring about human redemption and then showed the necessary mechanisms driving history in that direction. The French contributed two notable exemplars. Voltaire explained history as a movement away from ignorance and superstition toward rationality and enlightenment. Somewhat similarly, in a work entitled *Sketch for a Historical Picture of the Progress of the Human Mind*, Marie Jean Antoine Condorcet early in the 1790s anticipated idyllic conditions at the end of ten developing, historic stages, eventuating in "The abolition of inequality between nations, the progress of equality within each nation, and the true perfections of mankind."

German thinkers also harbored progressive expectations. Immanuel Kant's *Idea for a Universal History from a Cosmopolitan Point of View* produced an important statement. Published in 1784, a decade before Condorcet's *Sketch*, this essay affirmed freedom as the purpose of history. According to Kant's argument, "the history of mankind viewed as a whole, can be regarded as the realization of a hidden plan of nature" to bring about the requisite political conditions both within states and among them so that "all the capacities implanted by her [nature] in mankind can be fully developed."[2] Though somewhat vague about the mechanisms of attainment, Kant envisioned the creation of an international league of nations as the prime institutional means by which to achieve order and rationality within the world under law.

Though never a central preoccupation for Kant, history, according to the nine propositions set forth in this short piece, moved inexorably toward the goal of liberty as a consequence of what he called "unsocial sociability." This energizing principle, really the driving force of progress, in many ways the equivalent of Hegel's later idea, "the cunning of reason," expressed a favorite notion among Enlightenment philosophers, who liked to think that the pursuit of private interests automatically would translate into advancements for the collective good. Kant intended the term to characterize the devices employed by nature "to bring about the development of all the capacities implanted in men." Through the use of it, he referred to "their mutal antagonism in society," that is, as he explained in the fourth proposition, "their tendency to enter into society, conjoined, however, with an accompanying resistance which continually threatens to dissolve this society."[3]

To put this central claim another way, the conflicting and contradictory human needs to live together and to assert individual integrity paradoxically provided the momentum to move ahead. Typical of his time, Kant regarded stories about the past as consisting of lamentable tales of chaos, woe, disorder, and suffering. But in spite of the dismal record, he held forth better hopes for the future and intended his essay to hasten them along. In all likelihood, he wanted his writing to function as a kind of self-fulfilling prophecy as suggested by his choice of language in the eighth proposition. As the words reveal, he made no claim that history is "the realization of a hidden plan of nature," but rather said that it "may be regarded" as such. By implanting the thought, he hoped to encourage rational beings to act upon it as though it were true and thereby to attain the larger aim.

Another prophetic version appeared in the works of Karl Marx and his long-time collaborator, Friedrich Engels. Also linear and progressive in conception, their writings had an enduring and provocative effect. Unlike Kant, who only dabbled in history, Marx and Engels devoted a large portion of their lives to the endeavor and from it fashioned a theory of change and revolution. Characteristically, they too devised a vision of a more perfect future, but unlike utopians, visionaries, and metaphysicians, for whom they felt scorn, Marx and Engels believed that their analyses and predictions originated not in a priori imaginings but in a true comprehension of reality and a scientific understanding of historical development.

Their distinctive approach centered on crucial issues. First, Marx and Engels regarded the material world as the proper object of study, particularly the technologies of production and the economic, political, and social relationships generated by them. For Marx and Engels, the ownership of the means of production, that is, the tools, the lands, and the factories, established the main determinant of everything else. Second, they focused attention on a conception of class, defined in economic terms as ownership of property or lack thereof. As affirmed in *The Communist Manifesto*, the thesis held that "all history has been a history of class struggles, of struggles between exploited and exploiting, between dominated and dominating classes at various stages of social development." Last, they articulated an explanation of historical advancement and also of the mechanisms producing it.

Also Eurocentric in orientation, the Marx and Engels formulation resembled Hegel's in some respects. It depicted a historical movement from east to west and also an expansion of the levels of human consciousness. Marx, of course, reversed Hegel's view by vesting reality in the material rather than in the ideal world, but he accepted the dialectic as defining the logic and means of change. The ensuing term, dialectical materialism, put a label on Marx's assumptions about both. His subsequent division of the past into four great epochs had less to do with actual time periods than with his understanding of the various sets of human relationships in different systems of economic production. In some cases, they might even coexist.

The over-arching pattern discerned by Marx entailed the persistency of exploitation throughout all of history. It always pitted the "haves," the wealthy and the powerful, against the "have-nots," the poor and the vulnerable. It also featured systematic and institutional methods amounting to thievery by which the former relieved the latter of the just rewards of their labor. For Marx, the amount of work invested in anything established the true value, but under the prevailing conditions, the members of the labor force never received their just recompense. Instead, the ruling class appropriated the riches to its own use.

Marx posited the existence in the beginning of time of an age of primitive communism, marked by Garden of Eden-like qualities of simplicity, equality, and communal sharing, but then the Marxian equivalent of original sin, the advent of private property, shattered the idyll. Subsequently in Asia, the age of oriental despotism also featured an absence of private property and concentrations of population in small, dispersed, isolated communities under absolute rulers. Marx regarded such forms as dead, unchanging artifacts, receptive only to developments imposed from the outside. His real interest centered on Greece, Rome, and western Europe from which he derived a model marking the progression from the ancient to the medieval to the modern world. In ancient times, the class struggle took place in city states and aligned slaveowners against slaves. During the Middle Ages, the competition took the form of feudalism and divided lords and serfs over the ownership of land. In Marx's own day, the capitalist system set the owners of the means of production, the bourgeoisie, against the wage laborers, the proletarians, in the culminating conflict of interest and aspiration. According to Marx's calculations, the ensuing great revolution would overcome injustice and oppression through the abolition of private property and eventually would usher humankind into the new age of freedom and equality under the communist system in which all people would distribute fairly the rewards of their labor.

Marx's depictions of the past and his expectations for the

future provoked storms of criticism. Many scholarly works have claimed that his understanding of the ancient and medieval worlds suffered from fundamental flaws and that his other appraisals drastically overrated the revolutionary potential of modern times. Yet, his vision has retained the power to captivate, and many authorities agree that Marx knew more about the workings of nineteenth-century capitalism than anyone. Moreover, his insistence upon the need to study technology as an impetus toward transformation and change conveys an essential insight.

To simplify the general line of argument substantially, Marx asserted that the dominant, ruling class in any given epoch, that is, the males in charge of the technologies of production, also obtained control of political, legal, and social structures in order to safeguard their positions and interests. According to him, government officials in effect made up the board of directors of the economic ruling class. In similar fashion, legal arrangements, the social structure, and most other forms of organized human intercourse reflected the distribution of economic power. At the same time, the dynamism of historical imperatives threatened to unwork the status quo. Again and again in human history, the development of new technologies destabilized existing relationships and produced upheavals, rendering the old structures obsolete. For example, when the owners of factories obtained economic power, the former ruling elites, the feudal lords, struggled to retain their control of the political, legal, and social systems, but the very logic of history had doomed them. Once triumphant, the bourgeoisie erected suitable forms by which to preserve and maintain ascendency.

For Marx, all historical eras gave rise to exploitation and class oppression but nevertheless possessed integrity as necessary steps in the march toward communism. The attainment of perfection could never ensue without the necessary, foregoing prerequisites. For Marx, the model achieved truth from empirical proof and also from the irrefutable force of its logic. For Marx's critics, in contrast, the prophetic faith in revolution perhaps held less interest than the holistic analysis of human affairs and the explanation of the processes of historical change. Some even agreed that no need exists to embrace all of Marx in order to profit from his explication of the workings of economic capitalism.

Old-fashioned, speculative philosophy of history lost some of its appeal in the twentieth century, in part because of the traditional, grandiose aims. In an age of highly specialized inquiry in which, according to a well-known quip, we learn more and more about less and less, hard-nosed observers put little faith in any attempt to account for the whole of the human past. Indeed, they regard such endeavors as silly or contemptible, idle enterprises based on pure imagination and incapable of proof or demonstration. Although many turned their attention to other things, the older versions never died out. Indeed, they lived on, oftentimes in somewhat modified form. Marxian legacies in amalgamation with the ideas of Vladimir I. Lenin retained an intellectual and revolutionary appeal in various circles around the world. Oswald Spengler and Arnold Toynbee composed universal and philosophical histories. Sigmund Freud produced a body of writing with significance for historians, and religious thinkers such as Reinhold Niebuhr invoked history and memories of St. Augustine.

In the twentieth century, Marxist ideology became the official version of the past in communist countries and purported to present hopes for a better future in others. In hybrid form with Leninism, it held special importance as a theory of international relations according to which modern capitalist nation states act in the world arena primarily for economic reasons. Seeking cheap resources, investment outlets for surplus capital, and new markets to relieve overproduction at home, the capitalist powers by this critique would always engage in a chronic struggle for profit and advantage, sometimes colliding in armed conflict, until the victory of revolution would sweep away all need for such behavior. During the 1960s a group of historians in the United States called "the New Left" challenged the orthodox accounts of American foreign relations by arguing that the Marxist-Leninist model held true. A vociferous debate over the origins of the Cold War particularly occasioned a polemical duel when radicals claimed that the expansive nature of United States capitalism had brought it about. In more recent times, the emergence of "dependency theory" as an explanation for the plight of the Third World has yielded similar intensity. Initially a product of Latin America and inspired by the teachings of Marx and Lenin, dependency theorists attributed the pervasiveness of poverty and underdevelopment in much of the world to the requirements of a global capitalist system designed to drain away wealth from the provincial and colonial regions to the advanced metropolis. By this analysis, the ascendency of a world capitalist economic system will always result in oppression and exploitation. The controversy persists unabated into the present time.

In a similarly deterministic but more pessimistic fashion, Oswald Spengler, a morose, brilliant, and eccentric German prophet from the era of the First World War, harbored no apocalyptic visions of salvation or progress. The title in English of this erudite, two-volume work set the tone. *The Decline of the West* resurrected the cyclical view of history and argued the case that no exits permitted any escape from the dictates of destiny. Spengler identified culture as the proper unit of historical study. Though somewhat vague in his definition, he claimed the existence of eight distinctive cultures over the course of human history. He called them the Indian, Babylonian, Chinese, Egyptian, Mexican, Arabic, Classical, and Western. The distinguishing feature, according to Spengler, consisted of "a single, singularly constituted soul." In a confusing, mystical flight, Spengler affirmed, "A Culture is born in the moment when a great soul awakens out of the proto-spirituality of ever-childish humanity, and detaches itself, a form from the formless, a bounded and mortal thing from the boundless and enduring." In elaboration, he asserted, "It blooms on the soil of an exactly-definable landscape, to which plant-wise it remains bound. It dies when this soul has actualized the full sum of its possibilities in the shape of peoples, languages, dogmas, arts, states, sciences, and reverts into proto-soul."[4]

Spengler further compounded the difficulties by regarding culture as the analogue of an organism in the biological world, fated to pass through the phases of birth, youth, age, and death. Sometimes as an alternative, he used the images of spring, summer, fall, and winter. Each one had a life span of about one thousand years and also displayed structural equivalencies in the development of the life cycle. Spengler proposed to study them through the use of comparative morphology. Through the use of such techniques, Spengler intended to move beyond shallow comparisons to profounder insights. By this method, he concluded, Napoleon and Alexander the Great represented morphological "contemporaries," in that both stood in the same phase of a declining culture.

Spengler distinguished among the various cultures on the basis of diverse conceptions of time and space. An expression of cultural determinism, Spengler's provocative and revealing approach claimed that each distinctive culture dictated a different standard for perceiving and measuring time and space. Each culture thereby obtained uniqueness but at the same time cut off and isolated the human beings in any one of them from all the others. Spengler put little stock in the possibility of meaningful cross-cultural transfers. Any such attempt could only result in inexplicability. To work out the ramifications of his case, he concentrated his attention on three cultures, namely the Classical, the Arabic, and the Western, each with "souls" characterized as Apollinian, Magian, and Faustian.

Spengler's discourses typically described the prevailing conceptions of time and space within a culture and then marshalled a parade of illustrations from the worlds of mathematics, drama, architecture, politics, economics, and so forth to substantiate his claims. According to him, the Classical or Apollinian soul most cherished the qualities of timelessness in the present and closed space and attained the ideals preeminently in nude statuary and Euclidean geometry. The Arabic or Magian soul in contrast organized impressions around a sense of cavernous space and time spans with clear beginnings and ends. As examples, Spengler cited the early Christian caves, Muslim mosques, and the Judeo-Christian linear designs of history. Finally, the Western or Faustian soul conceived of pure and limitless space and time stretching out into infinity, represented by differential calculus, museums, clocks, and presumably space shots to Mars.

Whatever the observable variations, according to Spengler, all cultures must share a common fate. Destiny requires that they run their life course and die. On this point, Spengler curiously used the language of astrology. In the German title of his work, *Der Untergang des Abendlandes*, the first word has astrological significance, suggesting the decline or setting of the sun. As noted by one commentator, Bruce Mazlish, Spengler's sense of inevitability required foreordained outcomes because they are "in our stars." Mazlish referred to Spengler as "an astrological historian" for whom "Real history is heavy with fate but free of laws." As Mazlish remarks, "we can *divine* the future but not reckon or calculate it."[5]

Such crankish oddities abound in Spengler's writing. Though Spengler possessed a doctoral degree, he never possessed an academic appointment and mainly carried out his endeavors as an independent scholar. Perhaps his status as an outsider, thus removed from the mainstream of intellectual life in the universities, accounted for some of his quirks. His contemptuous disdain for other scholars and his pretentious boasts of an originality never previously attained rubbed many readers the wrong way. But Spengler struck a mood in the confusing years after the First World War and signaled a dramatic shift away from the sentimental optimism of the Victorian age.

Arnold Toynbee, an Englishman and a contemporary of Spengler, also suffered the effects of historical trauma during the Great War. In order to render historical study more systematic and compelling, he inaugurated an ambitious project in 1920. Upon the completion in 1954, Toynbee's majestic, twelve-volume work surveyed the whole experience of humankind and set forth a series of general propositions by which to convey the meaning. *A Study of History* ran parallel to Spengler's book in some respects but diverged in others. Most notably, Toynbee developed a particular interest in the origins of civilization, and moreover, he claimed to arrive at conclusions while adhering to the requirements of English empiricism. Though Toynbee professed to eschew the a priori techniques of his German counterpart, his critics later flailed him for falling short of the aim, charging that he unwarrantly designated the Greek and Roman worlds as the models of all civilization and that on occasion he twisted the evidence to make it fit his preconceived theories.

The Toynbee approach sought to obtain high levels of generality by depicting "societies" or "civilizations" as the proper units of historical study. Toynbee preferred them to nations or periods, because they partook of a more intelligible, larger whole. Though vague definitions caused problems for him and discrepancies in usage disorganized his count, Toynbee detected nineteen or twenty-one civilizations during the preceding six thousand years. Five still lived, our own Western civilization and also the Orthodox Christian, the Islamic, the Hindu, and the Far Eastern. The nonliving included the Hellenic, Syriac, Indic, Minoan, Sumeric, Hittite, Babylonic, Andean, Mexic, Yucatec, Mayan, and Egyptiac. Finally, two sets of "fossilized" relics maintained remnants in the present. These included Jews and Buddhists.

Toynbee reasoned that a majority of civilizations sprang from a predecessor but that some emerged directly from primitive life. The problem of the genesis especially intrigued him. He rejected race and environment as inadequate explanations and found the key in his conception of "challenge and response." As observed by D. C. Somervell, who abridged Toynbee's mighty work into two volumes:

> A survey of the great myths in which the wisdom of the human race is enshrined suggests the possibility that man achieves civilization, not as a result of superior biological endowment or geographical environment, but as a response to a challenge in a situation of special difficulty which rouses him to make a hitherto unprecedented effort.[6]

By Toynbee's analysis, civilizations took shape when demanding challenges brought forth creative human responses. Toynbee emphasized the importance of "the golden mean." Too severe a challenge could overwhelm all manner of reactions. Similarly, too bland a challenge would obviate any need to generate one. As examples, Toynbee noted that Viking emigrants from Norway succeeded in Iceland but collapsed in Greenland. Massachusetts, a tougher challenge than the South, elicited "a better response," but arctic regions proved too much.

Growth within a civilization, according to Toynbee, occurred when successive challenges met with creative responses. The responsibility for devising them resided with "creative minorities," that is, small groups of talented people who inspired others to follow through the force of their example. The ensuing process, dubbed *mimesis* by Toynbee, impelled imitation by the masses of their natural leaders. When mimesis broke down, or the creativity of the minority evaporated, then "a time of troubles" began and the society experienced the possibility of a breakdown.

Toynbee rejected strictly deterministic solutions to the problem. No inevitabilities required the disintegration of societies. Nevertheless, his scrutiny of the process showed that the transformation of creative minorities into merely dominant minorities and the emergence of an internal proletariat, which felt "in" but not "of" the society, signaled dangerous, impending schisms. Though spiritual resources always have done a great deal to determine the outcomes, Toynbee claimed that social decomposition came about not in a uniform fashion but by an alternation of "routs" and "rallies," his terms for defeats and recoveries. As Toynbee explained, the establishment of a universal state, such as the Roman Empire, represented a rally after a rout in a time of troubles, and the dissolution of a universal state indicated the final rout. Somewhat oddly, Toynbee claimed to have detected a rhythm in the process, consisting of three and a half beats: "rout-rally-rout-rally-rout-rally-rout." Claiming that his investigations into several extinct societies confirmed the pattern, he wondered about its applicability to his own civilization. In the abridgment, Somervell remarked, "As differentiation is the mark of growth, so standardization is the mark of disintegration."

In the last four volumes, religious themes became the dominant motif. Though convinced that his own civilization had advanced far into a time of troubles, thereby heralding the possibility of a decline and fall, he retained some hope that spiritual rebirth might halt the descent through the affirmation of a more creative alternative. In this philosophy of history, no deterministic laws fated any particular outcome. Everything depended upon the human actors, and religious conviction would function as the prime arbiter. In the concluding volumes, Toynbee emerged more as a religious thinker than as a historian.

Another quite different approach to philosophy of history emerged early in the twentieth century in the writings of Sigmund Freud, the founder of psychoanalysis. Though not primarily a historian, Freud developed an approach to the study of human behavior with immense implications for history. Rather than concentrate his attention on large collectivities, such as "cultures" or "civilization," he worked as a clinician with individual patients and then extended his findings and insights to the larger domain of humankind. Though critics have insisted that he misinterpreted the sexually repressed mores of turn-of-the-century Vienna for universal characteristics, his apologists replied, nevertheless, that he had uncovered hidden fundamentals at the core of human nature.

Freud discovered the motive force of individual behavior in the psyche, that is, the internal mental life of all persons. In order to comprehend the function of the governing mechanisms, he divided the mind into two parts, the conscious and the unconscious. The former, situated in the brain, produced deliberate and calculated thoughts and acts intended to achieve certain desires and designated goals and ends. The latter, the more remote and difficult to comprehend, by its very nature, lacked any awareness of itself but possessed, nevertheless, the key to the entire mystery.

To make possible an understanding of an otherwise impenetrable phenomenon, Freud posited the existence of three functions. First, the *id*, the oldest of the psychical provinces, consisted of everything inherited, that is, the instincts, a consuming cauldron of drives, desires, and cravings, seeking expression in forms unknown to the possessor. Second, the *ego* mediated between the demands of the id and the external world and defined the faces of reality. To an extent, it determined what kinds of satisfaction existed as possible by seeking after pleasure and avoiding pain. The third, the *superego*, perpetuated the experiences of childhood by prolonging into the present what Freud termed "this parental influence," including "in its operation not only the personalities of the actual parents but also the family, racial, and national traditions handed on through them, as well as the demands of the immediate social milieu which they represent." In Freud's view, these manifold considerations performed as something akin to a conscience. As Freud further explained, "the id and the super-ego have one thing in common: they both represent the influence of the past—the id the influence of heredity, the super-ego the influence, essentially, of what is taken over from other people—whereas the ego is principally determined by the individual's own experience, that is by accidental and contemporary events."[7]

The interaction and competition among these three components accounted for the unconscious mental life of all persons. Freud ranked the experiences of childhood as central. The relationships with parents and siblings and other such events, often long forgotten but still stored away in the recesses of the psyche, determined adult behavior in ways usually unrecognized by the actor. Mild disturbances, obsessions, compulsions, and the like, called "neuroses," required treatment through psychoanalysis, a technique administered by a trained practitioner to make the patient fully conscious of hitherto unconscious material. Through the ensuing confrontation between the known and the unknown, a resolution of the remembered events with the unconscious traces should result in a cure. For drastically disarranged mental states called "psychoses," Freud had more doubt about the possibilities of effecting much improvement.

Many of Freud's favorite phrases and ideas have become part of the jargon of the twentieth century. These include repression, projection, wish fulfillment, the analysis of dreams, the Freudian slip, and the Oedipus complex. Indeed, his emphasis on human sexuality as the source of the libidinal energy driving human beings has passed into current folklore. In contrast, his speculative writings on historical subjects have attracted less general attention but at the same time have spurred a consuming interest in some quarters among specialists who hope to achieve deeper levels of understanding through the use of psychoanalytical history.

Freud's books, *Totem and Taboo* (1912–13), *Civilization and Its Discontents* (1930), and *Moses and Monotheism* (1939) set forth suggestive models. Each concerned an analysis of psychical events derived from clinical observations and relationships with the external world. As Freud explained, he wanted to study "the interactions between human nature, cultural development and the precipitates of primaeval experiences." Religion figured prominently in these considerations and reflected "the dynamic conflicts between the ego, the id, and the super-ego, which psychoanalysis studies in the individual."

In *Totem and Taboo*, Freud combined some of the theoretical reflections of Charles Darwin with his own clinical observations. Freud borrowed a hypothesis holding that human beings originally lived in small hordes, each under the absolute rule of an older male who took all the females for himself and brutalized and killed the younger men, including his own sons. Such patriarchal oppression came to an end when the sons collectively rose in rebellion, destroyed the father-tyrant, and consumed his flesh. Consequently, they constituted a totemistic brother clan in which taboos renounced sexual contact with the women in favor of exonamy and, moreover, ambivalence over the killing of the father resulted in the adoption of a totem. Usually an animal, it stood for their ancestor and spirit protector. They would allow no one to hurt it, except in a ritual once a year when the whole clan feasted upon the creature. According to Freud, "it was the solemn repetition of the father-murder in which social order, moral laws, and religion had had their beginnings." For Freud, the actuality of any such event held little consequence. Whether real or fantasy, the psychic impulse to kill the father would have had the same outcomes.

Similar characteristics marked *Civilization and Its Discontents* and *Moses and Monotheism*. In the former, Freud engaged in the speculative philosophy of history by musing about the direction of historical movement, particularly the transition from the religious to the scientific stage in human existence. Though ethnically Jewish, Freud never embraced the religious traditions of his heritage and, indeed, subjected them to merciless criticism. All religions he classified as mass delusions, unhappy and misguided efforts to substitute a wish fulfillment for an unbearable reality. In this work, he called upon human beings to give up childish responses and stoically to face the truths. In *Moses and Monotheism*, he elaborated upon a related theme, claiming provocatively that Moses, actually an Egyptian, "chose" the Jews, led them out of bondage, and shaped them into a holy nation. But the Jews, reacting against the Mosaic laws, murdered their leader and thereby reenacted the tribal horde experience with similar consequences, except in this instance a monotheistic conception of deity replaced the totem.

Such inferences outraged more orthodox thinkers and have retained a capacity to disturb until the present day. Indeed, an animated controversy has drawn the line with the champions on one side, insisting that psychoanalytical history can arrive at more profound appraisals of character, and critics on the other who have rejected the approach merely as a body of theory, neither proven or provable. Perhaps most tellingly, the skeptics have depicted the techniques of Freud and his disciples primarily as therapies for which the only test of truth is practical. Does the patient get any better? If so, the practitioner has chosen the proper one and has arrived at a kind of truth. Unhappily, the historians employing psychoanalytical methods have no such means of verification and can only wallow in speculative ruminations. Whether such inferential leaps obscure the subject or advance understanding remains an issue under fierce debate.

In conclusion, a final example will illustrate the persistency of fundamental issues in the discussion over the speculative philosophy of history. In the United States during the 1940s, Reinhold Niebuhr attained formidable rank as a theologian, philosopher, and political commentator. A "neo-orthodox" Protestant in his religious commitments, Niebuhr set forth new versions of the Christian view but operated steadfastly within the larger traditions of St. Augustine.

For Niebuhr, neither the classical nor the modern secular renditions of philosophy of history held much merit. The Greek cyclical idea deprived history of real meaning by denying the significance of particular events. By submerging them in recurring patterns, according to Niebuhr, the Greeks denied to themselves the capacity to act creatively within the context of particular instances and sought to escape into an abstract world of pure thought, thereby transcending history altogether. In contrary fashion, the Western, linear, progressive construction of history put undue faith in the creative powers of human beings and found in history itself the means of increasing control over nature, the physical well-being of people, the democratization of society, and other such presumably desirable states of affairs. In other words, the very course of historical development in itself would result in human betterment through the enhancement of freedom and good.

Though Niebuhr held the modern version in higher regard than the classical, because it found meaning in the details of historical experience, he also believed that it suffered from a serious error by assuming a necessary connection between the growth of human freedom and good. For Niebuhr, no such relationship had ever existed. Indeed, to the contrary, the growth of human freedom and power simply "enlarges the scope of human problems" by giving greater opportunity for the expression of "egoistic desires and impulses." As examples, he noted that

> Modern industrial society dissolvèd ancient forms of political authoritarianism; but the tyrannies which grew on its soil proved more brutal and vexatious than the old ones. The inequalities rooted in landed property were levelled. But the more dynamic inequalities of a technical society became more perilous to the community than the more static forms of uneven power. The achievement of individual liberty was one of the genuine advances of bourgeois society. But this society also created atomic individuals, who, freed of the disciplines of the older organic communities, were lost in the mass; and became the prey of demagogues and charlatans who transmuted their individual anxieties and resentments into collective political power of demonic fury.[8]

Such basic Niebuhr showed his compelling taste for irony and paradox. Things seldom have turned out as people intended. Nevertheless, according to him, Christian conceptions of history guarded against naively optimistic expectations by espousing a more realistic and valid understanding of human nature. Christians knew the reality of original sin. This condition meant that, in the course of time, human beings could neither achieve their purposes consistently nor utilize their freedom and power to attain good. Nevertheless, Niebuhr insisted that the sovereignty of God presided over the whole of the human experience. As Niebuhr put it, "All historical destinies are under the dominion of a single divine sovereignty" which provides a "general frame of meaning for historical events." Niebuhr denied, moreover, that either reason or evidence could lend confirmation to any such claim. Only Christian faith could result in the certainty that "history is potentially and ultimately one story."

Niebuhr strove to work out the implications of his position in two important works during the 1940s, *The Nature and Destiny of Man* and *Faith and History*. These stood as Christian rejections of modern secularists, such as Marx and Freud, and returned the ancient debate to stark essentials. Niebuhr insisted upon meaning in history but found the ultimate source of meaning outside of history. Similarly, by rejecting the feasibility of rational or empirical demonstration, he removed the arbiter of debate from history to theology. Faith would have to decide. By such means, he closed the circle with St. Augustine.

RECOMMENDED READINGS

A compilation is available in Ronald H. Nash (ed.), *Ideas of History*, Vol. 1: *Speculative Approaches to History* (New York: E.P. Dutton, 1969). Bruce Mazlish, *The Riddle of History, The Great Speculators from Vico to Freud* (Minera Press, 1966) provides good introductory essays. Additional commentary exists in Frank E. Manuel, *Shapes of Philosophical History* (Stanford University Press, 1965); Karl Löwith, *Meaning in History, The Theological Implications of Philosophy of History* (University of Chicago Press, 1949); and Peter Munz, *The Shapes of Time, A New Look at the Philosophy of History* (Middletown, CT: Wesleyan University Press, 1977). In *The Poverty of Historicism* (New York: Harper & Row, Publishers, Inc., 1957), Karl R. Popper argues against the possibility of predicting the future. Introductions to Marx include David McLellan, *Karl Marx, His Life and Thought* (New York: Harper & Row, Publishers, Inc., 1973); M.M. Bober, *Karl Marx's Interpretation of History* (2d ed. rev.; New York: W.W. Norton & Co., 1965); Melvin Miller Rader, *Marx's Interpretation of History* (New York: Oxford University Press, 1979), and Gerald Allan Cohen, *Karl Marx's Theory of History; A Defense* (Princeton University Press, 1978).

Useful specialized works incorporate Isaiah Berlin, *Vico and Herder, Two Studies in the History of Ideas* (New York: The Viking Press, 1976); William A. Galston, *Kant and the Problem of History* (University of Chicago Press, 1975); and Burleigh Taylor Wilkins, *Hegel's Philosophy of History* (Ithaca, NY: Cornell University Press, 1974). *The New Psychohistory* (New York: Psychohistory Press, 1975), edited by Lloyd de Mause, sets forth a ringing defense. More critical works by Jacques Barzun, *Clio and the Doctors, Psycho-History, Quanto-History & History* (University of Chicago Press, 1974) and David E. Stannard, *Shrinking*

History, On Freud and the Failure of Psychohistory (New York: Oxford University Press, 1980) raise important questions. Freud's own book, *An Outline of Psycho-Analysis*, translated by James Strachey (rev. ed.; New York: W.W. Norton & Co., Inc., 1969) provides a resume of his thought. Finally, *The Irony of American History* (New York: Charles Scribner's Sons, 1952), by Reinhold Niebuhr, sets forth some of the main themes.

5

PHILOSOPHY
OF HISTORY:
Analytical
Approaches

The development of analytical or critical philosophy of history in the twentieth century has affected a quite different set of concerns. This branch of intellectual inquiry has addressed issues in methodology and epistemology, centering particularly on the crucial question: On what grounds can historians reasonably demonstrate that they know what they claim? In other words, the verifiability of historical knowledge comes under review. In pursuit of this discussion, one kind of criteria measures the logic, rigor, and techniques of history against the prevailing forms in the natural sciences. Other items under consideration include the formal requirements of an explanation, the meaning of causation, and the role of objectivity. Such issues have posed persistent problems and have elicted an ongoing debate nowhere near resolution. Nevertheless, they have vital significance in comprehending the theoretical dimensions of the discipline.

The main lines of division go back at least into the seventeenth and eighteenth centuries, when Francis Bacon, René

Descartes, and other scientists tried to establish more reliable means for studying the natural world. They and their successors favored mathematical formulations for testing and expressing their findings, and also they liked high levels of generality by which to affirm statements of invariable relationship. This latter phrase refers to recurring uniformities and regularities among phenomena. In rudimentary terms, they took the following form. Whenever some combination of prior conditions existed in a particular instance, then a certain and predictable kind of outcome necessarily would follow. To utilize an example from nature, whenever fire heated water to the proper temperature and in the right circumstances, then the ensuing consequence would produce steam.

Although the actual statements of invariable relationship became more intricate as the problems under observation grew more complex, the formal logical structure underlying them remained more or less constant, and for many natural scientists, the extent to which they could express them in mathematical terms, the better. Ordinarily the formal logic connected the anterior or preceding conditions to the observable outcome by means of affirming a general, empirical law. It said simply, after repeated verifications through experimentation, that whenever those kinds of conditions came into being, the same outcome or consequence would take place. Put crudely in categories more congenial to historians, the invocation of a general law meant that, given the same causes, very similar effects almost surely would occur.

Giambattista Vico in the eighteenth century repudiated such methods as inappropriate to historical studies, but nevertheless, the ascendency of scientific models overwhelmed his objections and resulted in a flatly dogmatic assertion. It held that all forms of knowledge must conform with the methods and techniques of natural science or else forfeit any rightful claim to the status of knowledge. Those presumably inferior kinds of inquiry, those less demanding in logical form, or less mathematical in orientation, would have two choices: either measure up against the dictates of "the hard sciences" or suffer disparagement as pseudoscience. For serious students of human affairs, the dilemma conversely left open two courses of action. First, they could conform with the demand, emulate the natural sciences, and seek to present their findings as general statements of invariable relationship. Such endeavors later would lead to the developments of the social sciences. Second, they could insist upon the propriety and integrity of the traditional methods and techniques within their own fields of study, disclaiming any need to adhere to the models of the natural sciences. This latter affirmation sometimes depicted history as *sui generis*, that is, a class of learning unto itself.

The first response achieved special prominence in the middle of the nineteenth century with the advent of positivism. Primarily the work of Auguste Comte, a Frenchman, but subsequently embraced and elaborated upon by two Englishmen, Henry Thomas Buckle and John Stuart Mill, this body of thought sought to transform the study of human affairs into a more systematic kind of inquiry by endorsing the techniques of natural science. Rather than focus on unique or individual kinds of events, positivists proposed to concentrate attention on uniformities and similarities in the course of human affairs and then to locate the invariable relationships linking the same kinds of experiences. Rather than study the French Revolution, they would investigate the phenomenon of revolution. In addition, positivists presumed the existence of general laws governing the outcomes of activity in the human world, just as in the natural world, and they put their trust in their own intellectual capabilities to find them.

The works of Auguste Comte gave rise to sociology and established the essentials in ten mighty volumes, *Cours de philosophie positive* (six volumes, 1830–42) and *Système de politique positive* (four volumes, 1851–54). To an extent, Comte's philosophy grew from his personal experiences, some of them a bit bizarre. Indeed, he obtained central insights leading to the Law of the Three Stages, a fundamental tenet, while suffering one of his periodic bouts with madness. The victim of an unhappy home life as a child, Comte fled to study mathematics at the École Polytechnique in Paris, abandoned Roman Catholicism, and for a time came under the influence of the French utopian, Henri de Saint-Simon. Though degrees of instability and eccentricity always marked his personal life and sometimes found expression in unique forms of religiosity, for example, his commitment to a religion of humanity, he possessed the intellectual means, nevertheless, to set forth an entire philosophical system called positivism.

Comte began with the familiar division of the past into three parts, claiming as a basic premise that the human mind had developed historically through three stages. Straightforwardly Eurocentric, he intended to focus his attention on the "vanguard of the human race," by which he meant the inhabitants of Italy, France, England, Germany, and Spain. He also preferred to proceed at a high level of abstraction, espousing his wish to employ history, except for reasons of stark practicality, "without the names of men, or even of nations." One commentator, the historian Bruce Mazlish, explained that this "dehumanization of history" came about because Comte really took primary interest in the sciences and their development. Ranked in ascending order of difficulty according to Comte, these included mathematics, really nothing more than a logical tool, astronomy, physics, chemistry, physiology, now called biology, and at the pinnacle, sociology.

A basic proposition affirmed the Law of the Three Stages by which Comte showed that the evolution of the human mind took place in three steps. He derived many of his convictions from the histories of the foregoing bodies of knowledge. In each, the affirmations about reality progressed from the theological through the metaphysical to the positive stages. In the first, human beings saw the world as controlled by wills independent of their own but subject to propitiation or manipulation by prayer or magic. During the second, abstract forces, such as the requirements of nature or the will of the people, governed all things, and in the third, the positive and final stage, an understanding of the invariable relationships among phenomena would provide for the explanations.

Much in the fashion of various predecessors, Comte also presumed holistic relationships, affirming that each mental stage corresponded with other kinds of intellectual and institutional developments. The theological phase, for example, coexisted with military life and primitive slavery, the metaphysical with lawyers and attempts at creating governments based on

law, and the positive with industrialism. But Comte's complicated efforts at refinement and elaboration also allowed them to
overlap, so that mathematics, the simplest among the forms of
knowledge, could achieve positive status in the otherwise theological age. But sociology, the most complex of all as the science
of human beings, could attain the position only when all other
forms of learning had also progressed to such a rank.

In the positive stage, no need would exist to indulge in idle
speculation over first or final causes. Instead, positivist principles would concentrate full attention on knowable objects and
the elucidation of law-like regularities which affirmed invariable relationships among phenomena. These in turn would not
rely upon theological or metaphysical presumptions but upon
empirical observations from the real world. Once advanced up
the ladder of learning from mathematics to astronomy to physics to chemistry to biology, the positivist philosophy would result in the creation of a new human science, sociology, the
articulation of which would result in new kinds of understanding of the laws governing human behavior and consequently in
new certainties over how best to calculate the probable outcomes of deliberate human acts.

The declaration of a positivist philosophy by Comte, Buckle,
and Mill produced a strenuously adverse reaction. Usually called
"idealist" within the context of this debate, this school of thought
emanated from the writings of Wilhelm Dilthey, a German;
Benedetto Croce, an Italian; and Robin G. Collingwood, an Englishman. Collectively they argued that the analogy derived from
the natural sciences could not hold up under the test and that
the intricacies of doing history required quite different conceptual schemes. Late in the nineteenth century, Dilthey drew a
fundamental distinction between the natural sciences
(*Naturwissenschaften*) on the one hand and the human sciences
(*Geisteswissenschaften*) on the other. The practice of either one
called forth distinctive methodologies. For one thing, according
to Dilthey, the natural scientist sought regularities and uniformities in nature upon which to base generalizations, while the
historian dealt with unique, specific, and unrepeatable events
outside of nature.

Later on in the twentieth century, Croce, one of Italy's most important intellectuals, and Collingwood, an Oxford philosopher, historian, and archeologist, elaborated upon the idealist conception of history. Croce emphasized the all-important consideration that historians existed in the present. For the past to take on vitality and meaning, historians had to make it come alive with relevance by rethinking it in the mind. For this reason, Croce drew the conclusion that "All history is contemporary history." Collingwood, often crotchety but seldom eccentric, dedicated his life to setting forth the fullest and most enduring statement of the idealist position. His great book, *The Idea of History*, published posthumously after his death in 1943 at age fifty-two, still ranks among the most significant books on the philosophy of history ever published in the English language.

To begin, Collingwood described history as "the science of human nature," the aim of which aspired to "self-knowledge." For him, the deceptive simplicity of that statement meant that the proper object of historical study centered on the human mind, or more properly, the activities of the human mind, and further that the appropriate means of investigation required "the methods of history." To employ a different phraseology, historians learned about mind by comprehending what mind had done. Collingwood then discussed the disparity between the methods of the natural sciences and the human sciences and explained why the former had no analogous relationships with the latter. "The historian, investigating any event in the past, makes a distinction between what may be called the outside and the inside of an event." With these words, he established the very crux of his message.

By the "outside of an event," Collingwood meant "everything belonging to it which can be described in terms of bodies and their movements: the passage of Caesar, accompanied by certain men, across a river called the Rubicon at one date, or the spilling of his blood on the floor of the senate-house at another." By "the inside of an event," he referred to that "which can only be described in terms of thought: Caesar's defiance of Republican law, or the clash of constitutional policy between himself and his assassins." To highlight the difference, Collingwood conceded that "The historian is never concerned with either of these to the exclusion of the other." As he explained, the work of the historian "may begin by discovering the outside of an event, but it can never end there; he must always remember that the event was an action, and that his main task is to think himself into this action, to discern the thought of his agent."

Collingwood then set forth another main thesis. "In the case of nature, this distinction between the outside and the inside of an event does not arise." He claimed, "The events of nature are merely events, not the acts of agents whose thought the scientist endeavors to trace." In other words what takes place in nature manifests no inner life. "To the scientist, nature is always, and merely a 'phenomenon,' . . . a spectacle presented to his intelligent observation." For the historian, in contrast, "the events of history are never mere phenomena, never mere spectacles for contemplation, but things which the historian looks, not at, but through, to discern the thought within them."

Collingwood then summarized his point by putting it still another way. In seeking to penetrate "to the inside of events and detecting the thought which they express, the historian is doing something which the scientist need not and cannot do." Nevertheless, the methodological prescription for attaining the goal produced great problems. How shall the historian make certain of the thoughts acted out by human beings? According to Collingwood, "There is only one way in which it can be done: by re-thinking them in his own mind." Or, as he affirmed in another place, "The history of thought, and therefore all history, is the re-enactment of past thought in the historian's own mind." This creative and critical experience "reveals to the historian the powers of his own mind" and hence of all minds, resulting in a fuller appreciation of human nature as revealed by the mind while working upon actual experience.[1]

For positivist critics, the idealist position partook of fantasy, mysticism, and self-deception. It called for explanations based on the operation of an unobservable entity called mind and required empathetic leaps into the heads of historical actors, all of which depended upon faith, not science. Under the criteria of the idealist alternative, the historian would know only when knowledge became knowable, and no other test would set forth a standard. For Comte's disciples, this premise lacked methodological integrity, defied verifiability through the use of evidence or observation, and belonged properly in the meta-physical stage of human development.

For Collingwood and his followers, in contrast, the appropriate rejoinders put down positivist charges on several grounds. First, idealists insisted that they possessed methodological integrity, claiming that their approach, if carried out prudently and rigorously, indeed, allowed for verifiable insights into the workings of mind. Through the correct use of documentary evidence, that is, letters, diaries, and the like, the historian could make legitimate inferences. Second, they regarded the role of active, critical thought about the past as significant. Only by such means could they bring life into history. Collingwood denounced dullish recitations of mere facts based on earlier authorities as "scissors-and-paste history." Finally, Collingwood had no use for the positivist emphasis upon the need for generalization. As he remarked in an important passage, "If, by historical thinking, we already understand how and why Napoleon established his ascendency in Revolutionary France, nothing is added to our understanding of that process by the statement (however true) that similar things have happened elsewhere." To reenforce the point, he added, "It is only when the particular fact cannot be understood by itself that such statements are of value."[2]

The controversy has become the dominant issue in philosophy of history and has permeated virtually all discussions of historical methodology. Divergent conceptions of explanation, causation, and objectivity bear directly on the matter and demonstrate additional aspects of the contemporary debate. These show further some of the complicated ramifications of historical thinking, thereby lending even more credence to an observation by a philosopher, W. H. Walsh: "The truth is that history is a much more difficult subject than it is often taken to be. . . ."

Another philosopher, Carl G. Hempel, established the main categories under discussion in 1942 with the publication of his seminal essay, "The Function of General Laws in History." In this presentation of the positivist case, Hempel argued that the explanatory forms of natural science indeed held true in the field of history. While defining a general law as "a statement of universal conditional form which is capable of being confirmed or disconfirmed by suitable empirical findings," he described the main function as "to connect events in patterns which are usually referred to as explanation and prediction." A quotation from Hempel will indicate the line of reasoning:

> The explanation of the occurrence of an event of some specific kind E at a certain place and time consists, as it is usually expressed, in indicating the causes or determining factors of E. Now the assertion that a set of events—say, of the kinds $C_1, C_2, \ldots, C_n$ —have caused the event to be explained, amounts to the statement that, according to certain general laws, a set of events of the kinds mentioned is regularly accompanied by an event of kind E. Thus, the scientific explanation of the event in question consists of
>
> > (1) a set of statements asserting the occurrence of events C_1, ... C_n at certain times and places,
> > (2) a set of universal hypotheses, such that
> > > (a) the statements of both groups are reasonably well confirmed by empirical evidence,
> > > (b) from the two groups of statements the sentence asserting the occurrence of event E can be logically deduced.

The ensuing analysis, though complicated and difficult to follow upon first encounter, arrived at an emphatic conclusion. It held that "The preceding considerations apply to explanation in history as well as in any other branch of empirical science." As Hempel affirmed, "Historical explanation, too, aims at showing that the event in question was not 'a matter of chance,' but was to be expected in view of certain antecedent or simultaneous conditions." Although in some instances "the universal hypotheses underlying a historical explanation are rather explicitly stated," Hempel conceded that, more characteristically, most "fail to include an explicit statement of the general regularities they presuppose" because "they are tacitly taken for granted" or they lack "sufficient precision." As he noted, affirmations about the movement of Dust Bowl farmers to California have assumed a "universal hypothesis" to the effect that populations in times of hardship "will tend to migrate to regions which offer better living conditions." By omitting the authority of a general law, historians have employed what Hempel called "an explanation sketch."[3] Though incomplete in detail, it had to adhere, nevertheless, to the logical requirements of the explanation form.

The controversy has consumed analytically minded philosophers of history ever since. Dubbed "the covering-law model" by William Dray, a critic, the Hempel position called upon historians to explain specific events by subsuming them under a general proposition. But dissenters found little point in doing so and argued against the exhortation on several grounds. First, they noted that the actual practices of historians seldom conformed with Hempel's prescription. Indeed, they argued, to follow it would render the entire exercise banal because of the absence of meaningful and agreed-upon general laws in the field of history. Either such generalizations do not exist or the lack of precision and specificity would deprive them of exactitude. Would anyone acquire a new and deeper understanding of the French Revolution by explaining that the French have always acted that way? Another criticism recalled Collingwood's critique. William Dray and others pointed out that the task of historians required them to make actions comprehensible within the context of the historical actors' own motives and aspirations. Once having accounted for the reasons for an act, the historian has nothing left to do.

In 1962, Hempel published another essay, seeking to meet the various attacks and to fend them off through tactical accommodations. This time he dealt more directly with the actual writings of historians and allowed them some greater degree of flexibility. According to his array of definitions, the "probabilistic-statistical form" of explanation maintained "to the effect that if certain specified conditions are realized, then an occurrence of such and such a kind will come about with such and such a statistical probability." The "elliptic or partial" explanation sketch presupposed the existence of general laws but failed to invoke them explicitly, and the "genetic explanation" accounted for changing conditions in history by linking them systematically to earlier ones. But in each instance, Hempel still insisted upon holding the historians to the requisites of what he now called "deductive-nomological explanation" forms, asserting in good positivist fashion that "the nature of understanding ... is basically the same in all areas of scientific inquiry." Indeed, he claimed to his own satisfaction to have demonstrated "the methodological unity of all empirical sciences."

The problem of understanding the nature of causation in history also intrudes upon the matter. The term "cause" has always produced perplexity. Historians usually have employed it in a kind of commonsense fashion so that ordinary readers have access to their discussions, but another effect sometimes results, much to the distress of logicians and philosophers, in ambiguity and imprecision. At least one authority, an Englishman, Michael Oakeshott, has recommended in favor of dropping the use of the term altogether.

To illustrate the manifold grounds for confusion, Aristotle once distinguished among four types of cause. They comprised the material, efficient, formal, and final. Among them, according to Ronald Nash, three have particular relevance to the study of history. An "efficient" cause consists of the prior events and conditions sufficient for the occurrence of the main event. For example, the cause of Lincoln's death was the bullet in his head fired by the assassin John Wilkes Booth. A "formal" cause in history accounts for an event through the location of "dispositional properties" needed to bring it about. As Nash notes, a window might break because glass is brittle, or Lincoln died because of the hatred of Northern sympathizers like Booth for the South. Last, a "final" cause accounts for an event by attributing it to the will or purpose of an agent. Still another cause of Lincoln's demise emanated from the desire of some Northerners to bring about more favorable conditions for the defeated Confederacy by eliminating a perceived enemy.

Advocates of the positivist and idealist schools also divide sharply over this question. The former, as adherents to the covering law model, must regard cause in the "efficient" sense as the most appropriate to history. Obviously, their preferred explanation form compels them to think of cause as a set of prior conditions. Moreover, as a kind of corollary, they usually conceive of cause as energy, a sort of mechanistic notion derived from physics. Just as the impact of the cue ball impels the eight ball into the side pocket in pool, so the combined effects of population pressures and land scarcities compelled people to move into the great American West. Idealists, in contrast, usually have understood the term in the sense of "final" cause. For them, it necessarily meant the will or intention of the historical actors. Purposeful action became the key. Once having ascertained the thought in Brutus' head upon stabbing Caesar, the historian could aspire to nothing else.

Still another difficulty has compounded the problem. While discussing causation, analytical philosophers often have established distinctions between "sufficient" and "necessary" conditions in accounting for the course of events. Sufficient conditions entail probabilities. A statement of them refers to the minimum requirements sufficient to allow an event to take place. The focus centers on likelihood rather than certainty. Necessary conditions in contrast involve inevitabilities. A statement of them would amount to a guarantee that a certain outcome would result. Since historians usually debate the causes of big and complicated events, such as the fall of the Roman Empire or the onset of the First World War, and they seldom can handle in their minds all the variables under consideration, the limitations of intellect dictate that they traffic only in sufficient conditions. Their narratives, rightly understood, always ponder probabilities, never inevitabilities.

The last issue under discussion in this chapter addresses the nature of objectivity. Often described as the most important and the most difficult methodological problem facing historians, objectivity requires that they withdraw personal preferences to the extent possible and seek to describe the subject under consideration as accurately as possible. But the degree to which historians have attained the goal and indeed can ever hope to do so has resulted in profound disagreements. Though the contours of this debate do not follow strictly the positivist-idealist cleavage, the two positions still suggest some of the magnitudes of difference.

The positivist argument once again would hold the historians to the standards of natural science, insisting that any other course would result in a diminution of rigor and believability. To demand anything less would disqualify history from the ranks of empirically-based, scientific disciplines and relegate it to the ranks of alchemy and phrenology. Yet, such criteria pose special impediments because they often presuppose a resort to laboratory techniques by which natural scientists seek to exclude variable factors and to exert maximum controls. Such procedures presume a large measure of repeatability. That is, anyone carrying out the experiment in the same way and under the same conditions ought to arrive at the same findings and conclusions.

For historians, this conception of objectivity presents two immediate obstacles. First, historians do not work in laboratories and can neither exclude confusing variables nor control conditions. For obvious reasons, their objects of study compel quite different approaches. Moreover, the idea of repeatability has no bearing. Indeed, a reverse kind of consideration has come into play. Historians have become notorious for their incapacity to arrive at solid agreements. Lock ten of them in a room (or a cell) with the same bodies of evidence, and they would in all likelihood arrive at ten divergent judgments.

Whether this aversion for consensus has reduced history to the status of pseudoscience has precipitated even more dissension. On the one hand, some historians, inclined toward a "relativist" persuasion, have argued that the measure of objectivity in history must always take on a different proportion than that in natural science. After all, historians traditionally have investigated all sorts of human matters charged with passion and emotion. By their very natures, they tend to divide observers and to set them at polar opposites. It is one thing to study pulleys and inclined planes, quite another the Second World War and the Holocaust. Moreover, the individual historians cannot always achieve detachment or indifference. The historian displays a bias through the mere choice of a subject and then compounds the problem by bringing other partialities to it derived from class, race, religion, or some other form of identification. To be sure, these need not represent gross bigotries. Nevertheless, they must affect the value judgments in the narratives and all attempts at evaluation. The historian also chooses his facts, picking them from an already incomplete roster of historical artifacts, enhancing thereby even further the chances for error.

An appropriate response might mute such misgivings by suggesting that, in fact, the liabilities do not exceed appreciably those of natural science. Similar hazards exist in both areas. For example, natural scientists also choose a subject for investigation and thus perhaps betray a bias acquired from life experiences. They also draw conclusions based on less than total knowledge, since no academic discipline aspires to recreate reality. Rather, they seek to obtain some insight into its nature. For such reasons, the predicaments of natural science run parallel in important respects with those of history.

Yet, in spite of it all, the historian does operate under a singular disadvantage. As the American historian Carl L. Becker observed, "In truth the actual past is gone; and the world of history is an intangible world, re-created in our minds."[4] In contrast with many forms of natural science, no actual "object" ever comes under observation. Instead, using the remnants of the past, the historian reconstructs history, employing statements of probability, not certainty, and subject always to the limitations of a point of view.

The angle of vision, or the point of view, renders historical narratives intelligible. Without them, historical artifacts would make no sense, and historical narratives would follow no coherent line of development. True, historians do not always agree, but divergent versions of the same events do not necessarily result in intellectual incompatibilities. Quite the contrary, disparate renditions may result in larger, complementary forms of understanding in which the one enriches and animates the other through the employment of different angles of vision. Theoretically, no limits restrict the number of "true" stories historians can tell about the past. Charles Beard, an American historian, always maintained that an economic interpretation of the Constitution in no way nullified political or ideological or institutional treatments. It merely added a new dimension.

Though the difficulty of sorting through the problem of objectivity no doubt will remain a source of perplexity and consternation, the historian Paul K. Conkin has suggested a wise way of moderating the dilemma. While properly refusing to abandon conceptions of truth and believability in historical narratives, he argued nevertheless that somewhat different standards of objectivity must apply than in the natural sciences. The rules of reason and logic, of course, must determine the narrative forms, and the historian, moreover, must possess sufficient public evidence from which to draw plausible and valid inferences. In a pithy passage, he affirmed his understanding of the core of the matter:

> After finding causes, describing complex wholes, or finally constructing stories and disciplining them to fit evidence, we may come to think of our conceptualized past as a fixed, preexisting structure, awaiting our discovery, rather than seeing history for what it is—an imaginative, time-conditioned, but evidenced human invention. But reality as experienced severely limits those concepts that intend to explain a part of reality and that intend to bring us into some harmonious, working relationship to that part. And it is in our dutiful respect for those limits that, in any discipline, we are objective.[5]

The foregoing presentation aims at resolving none of the theoretical issues in dispute but rather at providing some guidance in following the main lines of debate and also at sounding the alert to some of the methodological pitfalls. Upon serious encounter, the latter allow for no facile assumptions about much of anything. The ongoing controversy over the "positivist" and "idealist" models have shaped the discussion but do not imply the existence of two different kinds of reality. Rather, they suggest two divergent attempts at comprehending portions of the natural and the historical worlds. They also recall the distinction established by Sir Isaiah Berlin in an essay on Leo Tolstoy's philosophy of history. Entitled *The Hedgehog and the Fox*, the line comes from a piece of poetry by the Greek Archilochus which said: "The fox knows many things, but the hedgehog knows one big thing."

Berlin believed that those words marked one of the deepest differences among writers, thinkers, and human beings in general. As he explained in a kind of extended thesis statement:

> For there exists a great chasm between those, on the one side, who relate everything to a single central vision, one system less or more coherent or articulate, in terms of which they understand, think and feel—a single, universal, organizing principle in terms of which alone all that they are and say has significance—and, on the other side, those who pursue many ends, often unrelated and even contradictory, connected, if at all, only in some *de facto* way, for some psychological or physiological cause, related by no moral

or aesthetic principle; these last lead lives, perform acts, and entertain ideas that are centrifugal rather than centripetal, their thought is scattered or diffused, moving on many levels, seizing upon the essence of a vast variety of experiences and objects for what they are in themselves, without, consciously or unconsciously, seeking to fit them into, or exclude them from, any one unchanging, all-embracing, sometimes self-contradictory and incomplete, at times fanatical, unitary inner vision.

In an additional elaboration, he averred, "The first kind of intellectual and artistic personality belongs to the hedgehogs, the second to the foxes." Without insisting upon any kind of rigid classification, he suggested that Dante belonged to the first category, and Shakespeare to the second. Plato, Lucretius, Pascal, Hegel, Dostoevsky, Nietzsche, Ibsen, and Proust ranked as hedgehogs, and Herodotus, Aristotle, Montaigne, Erasmus, Molière, Goethe, Pushkin, Balzac, and Joyce were foxes. His observations pointed toward something fundamental and widespread. If they held credence, then perhaps the very existence of the "positivist-idealist" dichotomy partook of the universal and expressed a distinctive facet of the workings of mind. If so, it may brook no resolution.[6]

RECOMMENDED READINGS

Useful compilations are contained in Ronald H. Nash (ed.), *Ideas of History*, Vol. 2: *The Critical Philosophy of History* (New York: E.P. Dutton, 1969); Patrick Gardiner (ed.), *Theories of History* (New York: The Free Press, 1959); and William H. Dray (ed.), *Philosophical Analysis and History* (New York: Harper & Row, Publishers, Inc., 1966). William H. Dray, *Philosophy of History* (Englewood Cliffs, NJ: Prentice-Hall, Inc., 1964); W.H. Walsh, *Philosophy of History, An Introduction* (rev. ed.; New York: Harper & Row, Publishers, Inc., 1967); and the segment by Paul K. Conkin in Conkin and Roland N. Stromberg, *The Heritage and Challenge of History* (New York: Dodd, Mead & Company, 1971) provide additional materials. The main scholarly journal in this area is *History and Theory, Studies in the Philosophy of History*.

The debate over "the covering law" figures prominently in Arthur C. Danto, *Analytical Philosophy of History* (Cambridge: At the University Press, 1965); William Dray, *Laws and Explanation in History* (New York: Oxford University Press, 1957); Haskell Fain, *Between Philosophy and History, The Resurrection of Speculative Philosophy of History within the Analytic Tradition* (Princeton University Press, 1970); Leon J. Goldstein, *Historical Knowing* (Austin: University of Texas Press, 1976); J.H. Hexter, *Doing History* (Bloomington: Indiana University Press, 1971); and Morton White, *Foundations of Historical Knowledge* (New York: Harper & Row, Publishers, Inc., 1965). The two essays by Carl G. Hempel are reprinted in Nash and Gardiner. Representative writings by Carl Becker and Charles Beard are also in Nash. The emergence of an alternate view is examined by Georg G. Iggers, *The German Conception of History, The National Tradition of Historical Thought from Herder to the Present* (Middletown, CT: Wesleyan University Press, 1968). The premier statement of the "idealist" position, of course, is R.G. Collingwood, *The Idea of History* (New York: Oxford University Press, 1956). The concluding consideration comes from Isaiah Berlin, *The Hedgehog and the Fox* (A Mentor Book, 1957).

6

READING, WRITING, AND RESEARCH

Among the activities of historians, whether beginners or advanced, the responsibilities of reading, writing, and research hold special importance. As consumers and producers of scholarship, all students of history need to cultivate some particular kinds of skills. The reading requires certain powers of retention and synthesis, a capacity to work over large bodies of information and to establish a measure of intellectual possession. The writing calls for an ability to communicate clearly, in the case of history, in plain, jargon-free prose; and the research compels, among other things, orderly, systematic, and imaginative forms of inquiry. The degree to which historians can attain such capabilities will determine their successes and failures.

Historical writing falls into several categories. The textbook, the most pervasive on college campuses, contains general introductory accounts. It functions as a place to begin and as a sort of reference work. Others include narrowly conceived scholarly articles and monographs, broader synthetic works, and doctoral dissertations. Each has a standard format and a particular role to play. Within the guild, such types of writing bear the descriptive name, "secondary source." They are "secondary" in that historians have derived them from "primary sources," that is, documentary materials in archives and other such artifacts.

Doctoral dissertations serve as a launch pad for aspiring scholars. They come into being as a concluding experience at the end of graduate school and mark the advent of a professional career. In most cases, they focus on a small, even minuscule, topic, possibly something like the politics of the Whig party in Massachusetts during the war with Mexico. They almost always show the effects of pressure and deadlines. Among other things, they demonstrate a capacity to carry out research in primary sources and to write up the findings intelligibly. Completed dissertations affirm credentials of a sort but never have won over large audiences. For a variety of reasons, the readership has remained small and elite, mainly consisting of the examining committee and perhaps members of the immediate family.

Doctoral dissertations oftentimes turn into scholarly articles and monographs. Such outlets make possible the dissemination of the findings of specialized research and primarily address professional scholars. Periodicals such as *The Journal of American History* and *The American Historical Review*, among the most prestigious in the United States, publish essays of wide interest to historians generally, while others, such as *Diplomatic History* or *The Annals of Iowa*, center on themes and regions. A unique journal of some kind provides for discussion of practically every subfield. Consider as examples *Forest History, The Hispanic American Historical Review*, and *The Journal of Negro History*. Scholarly monographs, in contrast, usually appear as books and also aim primarily at other professional historians. Characteristically based on extensive work in archives, they explore a narrow subject in substantial detail and often argue an explicit thesis. In other words, they present the research findings within the context of a particular line of reasoning. Such publications have titles such as *Pan American Visions: Woodrow Wilson and Regional Integration in Latin America, 1913–1921*, or

The Great Anatolian Rebellion, 1000–1020/1591–1611. They also have comparatively small audiences.

Synthetic works in history often project more ambitious goals. Usually an outgrowth of the author's own specialized labors in the field, this kind of book also draws extensively on the research of others and seeks to treat the subject under discussion in broad and accessible terms. Such works invite the attention of intelligent, lay readers. As works of synthesis, these publications investigate a period or pursue a theme, while drawing on the whole body of pertinent scholarship in order to develop an interpretation. They have titles such as *Europe and the French Imperium, 1799–1814* or *In the Shadow of FDR, From Harry Truman to Ronald Reagan.*

The formal writing by students, though perhaps more modest in purpose and intention, comprises at least three types. These include the interpretive essay, the book review, and the research paper. Though specifications will vary, depending upon the wishes of individual instructors, some general pointers may serve a useful purpose. Again, the object is suggestive, not prescriptive. Some additional hints on clear writing will follow.

The interpretive essay calls for a reasoned response to a question of some sort, presented always with proper regard for the rules of logic and evidence. The exercise may take place in class as an essay examination or outside as a take-home project. In either instance, the fundamental expectation remains much the same. The student needs to develop a coherent and defensible line of thinking. A wise instructor will often prize encyclopedic but uncritical essays less than those characterized by a readiness to engage in analysis, generalization, and evaluation. Critical thinking, after all, is the point.

Most kinds of formal writing should begin with some sort of thesis statement. It affirms the main theme under consideration and provides an indication of how the narrative will unfold. The first paragraph of an interpretive essay should assert the thesis and then the ensuing segments should set forth the reasoning and the evidence. Suppose the question asked for a discussion of the extent to which United States imperial expansion at the turn of the twentieth century conformed with the

Marxist-Leninist explanation. Any conceivable answer would have to presume some knowledge of Marxism-Leninism; the thesis statement would set forth a claim either in favor of the proposition or against it, and then the main body of the presentation would round out the argument with specifics, details, examples, illustrations, and the like. Critical analysis should run throughout, and at the end, a conclusion should reaffirm the point of the thesis, now regarded as a working hypothesis vindicated by the preceding analysis.

A book review calls for an appraisal of a work published by a practicing writer. Readily accessible examples in history abound in scholarly journals, such as *The American Historical Review*. Though professional historians often dash them off routinely, the book review gives beginning students a special problem because they lack experience and familiarity with the other literature on the subject. At the minimum, this kind of writing should accomplish two aims. It should describe the contents of the book and also develop a critical evaluation.

When picking a book to review, students should make a selection based on intrinsic interest and appeal, if they have been given any range of choice. They should take brief reading notes as they proceed through the volume. This practice will help them keep track of the discussion and also will assist them in the writing. As a flexible rule of thumb, about one page of review for each one hundred pages of text will suffice. A full bibliographical citation should adorn the top of the page, followed by the commentary, presented in proper, idiomatic English with an appropriate regard for the rules of grammar and style.

Book reviews ordinarily incorporate two parts. The first provides a synthesis of the contents, conveying some idea of the book's subject and how the author addresses it. The second sets forth a critical evaluation and asks most essentially, does the book articulate believable, verifiable, and legitimate claims? In arriving at such judgments, the book reviewer needs to ask at least three additional questions. Are the lines of argument logical, coherent, and consistent? Does the evidence presented allow just grounds for making the inferences? Does the author have appropriate qualifications for writing such a book? In other words, the reviewer must look out for signs of bias, prejudice, distortion, and misrepresentation and then arrive at an appraisal accordingly.

The research paper, to be sure a stern and demanding test, often elicits a panicked response. Characteristically perceived in student circles as cruel and unusual punishment, and therefore of dubious constitutional standing, the opportunity to engage independently in research also holds forth opportunities, as long as the student proceeds under appropriate guidance and with some sense for the proper and feasible. Adherence to the following precautions should ease the emotional trauma and enhance the chance of taking part in an intellectual experience. Indeed, with genuine care and planning, some measure of personal satisfaction may even result.

First and foremost, the selection of a topic holds critical importance. Many students doom their own efforts at the outset through the choice of a hopelessly dull or unrealistic topic. For most undergraduates, a treatise on the origin of street names in Swink, Colorado, would not hold much promise. Moreover, students need to calculate whether they possess the appropriate library resources, the necessary specialized knowledge and skills, and the capacity to cover the subject adequately within the allowable space and time. Aspiring researchers should shy away from investigations of subjects such as "French Military Debacles since Napoleon" if they have no access to French works on military history, no knowledge of the French language, and no more than fifteen pages in which to carry out the assignment. As a general rule, students should assume that the more precisely they can define and circumscribe their subjects, so much the better. In other words, they need to establish rules of inclusion and exclusion and to center their efforts to the extent possible on a unified theme or purpose. Imagination always figures prominently by giving vitality to the enterprise.

Once having decided upon a workable topic, the novice scholar next needs to mount a systematic search for research materials. Merely going to the card catalogue and thumbing through the files can have undesirable effects. For one, the findings will reflect only the holdings of the library in question. A more demanding and effective procedure would direct the student to seek out as many pertinent articles, books, and other such sources as possible. The fulfillment of this aim usually requires consultation with bibliographical guides and aides. Examples include the *International Index to Periodicals*, the *Readers' Guide to Periodical Literature*, the *Handbook of Latin American Studies*, the *Harvard Guide to American History*, and the *Guide to American Foreign Relations since 1700*. Many other such bibliographical tools exist, and reference libraries, of course, can provide indispensable help.

The efficient conduct of research requires system, order, discipline, and good habits. Two inviolable rules underscore the point and the need for accuracy. First, always obtain complete bibliographical citations for all sources and inscribe them on 3x5 index cards for the files. Incomplete or erroneous information will always cause subsequent difficulties and delays. Second, take usable research notes, always indicating the location of the source and the pages from which the information comes. In the transcription, ordinarily only one idea should go on any single note, and paraphrase usually has more utility than direct quotation. The former practice facilitates the task of putting the material into one's own words. If a quote has special charm, then absolute accuracy must be used to retain it.

The mechanical process of taking notes can assume several forms. Some researchers like to write their findings in loose-leaf notebooks, dividing carefully the various entries so that, later, they can tear them into strips and organize them topically and chronologically for the writing. Another approach employs the use of 5x7 cards for easy shuffling, and in most recent times, word-processing techniques have become popular. Under any method, the need for systematic, orderly, and accurate thoroughness holds vital importance. To make the point another way, the researcher must carry out each task with painstaking completeness and do it the same way every time.

While writing the paper, the appropriate techniques for citing sources and invoking authorities have always given students problems. Whether to employ footnotes or endnotes depends upon the wishes of the instructor. In contrast, when compiling the notes of either sort, the writer must make certain judgments. As another rule of thumb, authors can assume that if an ordinary textbook on the subject would contain a piece of information, then no need exists to verify it in a note. On the other hand, if a well-informed reader probably would not know, then a note should appear. Another implication also exists. The rules of scholarship require the attribution of direct quotation and also of extended paraphrases. To do otherwise runs the risk of plagiarism. It amounts to the worst crime in academe and consists of taking someone else's words or ideas and passing them off as one's own without giving proper credit.

The actual writing should follow the rules of grammar and style and can achieve good results through the exercise of care, practice, and patience. Important hallmarks of effective historical writing consist of clarity, precision, and accessibility. In most cases, historians seek to write their narratives in plain English and to convey their thoughts exactly enough so that all interested readers can obtain the means of entry. The following suggestions, while hardly exhaustive, may provide some guidance.

First, the structure and organization of formal writing require planning and restraint. Reliance on metaphorical language, the free association of ideas, or other such devices has an important place in some kinds of literature but normally not in history. In this field, pointedness, simplicity, and directness establish the means of communication. Another characteristic features a compelling logical flow to move the reader readily from one point to the next.

Sentences and paragraphs make up the building blocks. Sentences contain at least a subject and a verb. The rules of formal writing forbid the use of any construction other than a complete sentence. Paragraphs begin with a topic sentence and articulate a single thought or a closely related group of thoughts. At the end, a transitional sentence should facilitate the movement from one paragraph to the next. Throughout, logic and reason should keep similar subjects and themes in close proximity.

Second, effective formal writing compels close adherence to the rules of grammar and style. At bare minimum, this injunction requires that subjects and verbs agree, that correct punctuation separate the various units of composition, that the consistent use of capital letters designate some proper nouns, that apostrophes indicate possession, and that ambiguous antecedents never appear. Large measures of confusion and ignorance surround such matters, compounded often by a cavalier attitude. Philistines pay little heed, demeaning the role of the capital, dismissing the apostrophe, and treating commas, periods, colons, semicolons, dashes, and hyphens as interchangeable. Such lawless and disrespectful practices run many risks, including the loss of communicative skills. Happily, good English grammar can provide a remedy.

Questions of style depend less upon rules than upon the dictates of good taste. They impinge upon a variety of concerns. For example, beginning writers should rely primarily upon short, declarative sentences. To be sure, they need to vary the structure and the rhythm, but they should never affect grandiloquence and never employ turgid, complex, pompous, or bombastic constructions. To express the same thought in positive form, they should present the main thought conspicuously in the main clause without undue embellishment and then let it go. A related concern focuses on excessive verbiage. To make the point in command form, eschew wordiness. Never use more words than necessary.

Tense, voice, and verb selection also bear on the matter. In most cases in historical writing, the simple past tense is most appropriate. By and large, inexperienced writers should favor the active voice more than the passive voice. This distinction emphasizes the importance of employing strong verbs. In the first, the subject acts upon the object directly and yields a lean, muscular sentence. "During visits to Washington, Colonel House influenced President Wilson's conduct of international statecraft." The following, in contrast, produces a backward and clumsy sentence in which the object acts upon the verb. "In the conduct of international statecraft, President Wilson was influenced by Colonel House during his visits to Washington." A particularly annoying example asserts, "The first trip to Boston will always be remembered by me."

Responsible students should avoid such abominations, unless perhaps they aspire to careers in bureaucracy, government, university administration, and certain aspects of law and insurance, in which cases studied ambiguity and indirection often takes on some measure of utility. Press secretaries for any high official could not dispense with phrases such as "it has been decided" or "it will be determined." These passive voice constructions present masterpieces of evasion which disallow any attempt to fix resonsibility. In contrast, the careful writer concerned with clarity will stick with strong, active verbs.

Skilled writers try to stay away from jargon, clichés, and euphemisms. To be sure, some kinds of subjects require the use of technical terms, but generally the principle still holds in favor of simple language. Some examples will illustrate the nature of the problem. After conducting a briefing at Colorado State University, an Air Force general actually expressed gratitude for the opportunity "to interface" with his audience. Computer terminology in this instance poses a threat. Similar fancy renditions might describe two professors talking as taking part in "information transfer." Clarity remains the aim. During the Watergate scandal, when President Nixon's press secretary, Ron Ziegler, characterized certain statements as "inoperative," he meant they were false. Later, when President Carter referred to the attempt to rescue the hostages in Iran as "an incomplete success," he meant that it had failed. Considerations of proper usage would also forbid the employment of words ending in "ize," such as "maximize' or "prioritize," and also the omnipresent "hopefully."

The following mangled sentences illustrate some additional errors to avoid. Though unintentionally comic in effect, they represent serious lapses of a most debilitating sort and demonstrate the need always to edit carefully.

Words Out of Proper Proximity:

"I saw a woman covered with blood holding her head in her hands."
"Babies have been known to hiccough before they are born. An unborn child has been heard as far as 25 feet away from its mother."
"Four Indian tribes are selling liquor on the reservation they purchased from out-of-state wholesalers."

"The exiles closely watched the overhead monitors, wondering whether the plane would be late again and sipping Cuban coffee."

"Victor Parker has the title role in 'Medea,' a 1948 adaption of Euripides' play about betrayal and revenge at Portland State University."

"Lloyd Hand is the son of a Danish sea captain who had to quit school at 13 when his father died to go to work."

"Late night viewers were surprised to see Johnny Carson show up following his recent vacation with a beard."

The Double Negative:

"Type A behavior frequently begins very early in childhood because inadequate amounts of unconditional love and affection are not given to youngsters by both of their parents.

The Mixed Metaphor:

"In several recently published studies, the importance of the disciplines in the arts, humanities, and social sciences has been recognized as pivotal in the academic fabric of the university."

Bureaucratic Gibberish:

"There have been several students come in and reduce their work-study awards this semester. With this extra money, I may be able to increase the allocation of students in your department that are running out of funds. There is no restriction on type of work-study, either need or merit, but I must be able to fit the increased award into the student's financial aid budget. Have your students that may be running out of money come and see me. Hopefully I can help them finish the school year on work-study and save you money."

Such stylistic atrocities provide just cause for sobriety and restraint, and they bear equally upon the writings of amateurs and professionals.

The process of composition and publication by professional historians takes place at the conclusion of extended efforts. For most, the research experience, alternately wearying and exhilarating, concentrates work in the "primary" sources. They come in various kinds. Some appear in published form as books, others as reproductions on microfilm or microcards, and many in archives and libraries as original manuscripts. The conduct of research sometimes requires travel to many locations, surely a significant fringe benefit. Knowing the location of collections and the means of access thus takes on substantial importance.

Published documents provide historians with a place to begin. Many governments print more or less systematially collections of documents, proceedings, and state papers. To an extent, these endeavors fulfill a propaganda purpose and establish "official" versions of events. Significant series such as the *Papers Relating to the Foreign Relations of the United States* and the *Congressional Record*, nevertheless, serve the interests of historians but sometimes induce a kind of adversarial relationship. Scholars have a vested stake in openness and accessibility of information, while officials typically put less stock in making public the whole truth. Another kind of published primary source consists of the personal papers of influential people, such as Thomas Jefferson, Theodore Roosevelt, or Booker T. Washington. Usually put out by university presses, these expensive and ambitious undertakings make available vital research material and usually exist among the holdings of good libraries.

Reproductions of microfilm and microcards, in contrast, make possible the dissemination of historical records at relatively low cost. Though sometimes cumbersome and annoying to use, such devices reduce travel costs and make possible the working through of documentary collections while staying at home with a mechanical reader. These efforts serve commendable purposes. The National Archives of the United States has benefited the community of scholars in all sorts of ways by putting out a host of microfilm collections, comprising many of the records of the Departments of State, War, Navy, Treasury, Justice, and Interior.

Manuscript collections constitute the raw material from which historians fashion their narratives, and they have a fixed location. The actual sites range from college libraries to local and regional historical societies to official governmental archives at all levels to presidential libraries. Most grant access to qualified researchers without undue harassment but also maintain security precautions against theft. In the Library of Congress and the National Archives in Washington, D.C., armed guards scrutinize the scholars at work and routinely search briefcases and book bags. Classification restrictions also present hazards for historians interested in the very recent past. The right to know does not always apply, especially in delicate areas concerning military and diplomatic affairs. Governments in England and western Europe by and large adhere to a fifty-year rule, allowing the records to become available after the elapse of time. In the United States, the authorities traditionally have tried to achieve declassification more rapidly, but have had problems in obtaining the goal.

The actual contents of archival collections consist of the original correspondence, memoranda, and diary entries exchanged and recorded by the historical actors. All such artifacts require special care and handling. The major repositories in the United States include the National Archives, the Library of Congress, and the various presidential libraries. The first contains the records of federal agencies, the second, the collected personal papers of notables, and the third, research materials on all of the presidencies since Herbert Hoover. The decentralization of archival resources in presidential libraries has caused logistical problems for scholars who need to travel about from one to another; nevertheless it has become the standard practice. Archival research in foreign countries affords special opportunities and hazards. Differences in language, culture, and custom all create potential impediments, yet for many historians the conduct of research in another country ranks among the most challenging and enriching of experiences. More often than not, the joys of pursuing the hunt provide compensation for the frustrations.

The foregoing discussion may suggest to some readers that an undergraduate experience with history ought properly to result in the professional practice of history. Such, of course, is not the case. The vast majority of history majors never become professors. Quite the contrary, they pursue an immense diversity of professions. Quite literally, they do everything under the sun, developing careers in teaching, law, the ministry, the military, business, government, and other such areas. Limited only by the extent of their own imaginations, they possess many transferable skills and can do many different things.

As students of the liberal arts, history majors possess many capabilities of which they are sometimes insufficiently aware. If they have cultivated their talents properly, they should emerge from their colleges and universities with a capacity to write and speak well, to think critically, to carry out research, and to make judgments about large and confusing masses of information. The attainment of literacy, always a laudable goal, should serve them well in all fields of endeavor. The ability to articulate thoughts clearly and to interpret them correctly will always establish an edge.

Though vocationalism and an obsessional concern for getting a job have pervaded higher education in the United States in recent times, those disciplines making up "the liberal arts" have played their traditional role by upholding the importance of creativity, understanding, and communication. Without them, nothing much can function properly. Various studies now suggest that the average twenty-one-year-old will switch careers no less than three times in a working lifetime. This prediction, even if only approximately true, should reaffirm the veracity of the liberal arts as a means of achieving adaptability. The knowledge never becomes obsolete. The following skills inventory underscores the point by heightening a sense of awareness. Though presented as a practical aid in obtaining employment, it also shows some of the assets derived from some exposure to the disciplines in the liberal arts.

SKILLS INVENTORY

Courtesy of Career Development Center, Gustavus Adolphus College.

Before you begin the interviewing process connected with a job hunt, it is important that you develop a strong knowledge base of your qualifications. Over the years, you have developed many transferable skills from coursework, extracurricular activities, and your total life experiences. If you've researched top-ics, written, edited, and presented a paper for a history class, you've used transferable skills. Your planning and coordinating of a spring banquet used transferables. These skills are not limited to any one academic discipline or knowledge area. They are found in courses and other experiences that are broadly applicable to many occupational areas. Students who can describe their skills to a prospective employer will present themselves as someone who knows how to apply their liberal education to a work environment.

The Skills Checklist

On the next four pages is a list of five broad skill areas which are developed in a liberal arts college. Below the general heading is a breakdown of more specific skills:

Instructions:

1. Review the list and check (✔) those skills which you feel you possess.
2. Review those skills you checked. In the column next to the checklist, mark your skills with a rating to indicate your ability in each area:
   1 = enough ability to get by, with help from others
   2 = some ability
   3 = definite, strong ability in this area
3. In the fourth column, "Evidence," record examples of situations and specific results that support your selections of skills you see as being rated "2" or "3"—for example, if you checked "Motivating," record the specific situation: "Teaching older women to swim involved constant motivation and encouragement."

There are likely many more skills than are listed on this worksheet. Explore your personal experiences in work, school, and your social environment. Be sure to record any additional skills you have developed.

SKILLS INVENTORY

SKILL	✔	RATING 1–3	EVIDENCE
COMMUNICATION: The skillful expression, transmission, and interpretation of knowledge and ideas.			COMMUNICATION:
Effective Speaking...........			
Concise Writing..............			
Attentive Listening			
Expression of Ideas..........			
Facilitating Group Discussion ..			
Appropriate Feedback........			
Negotiating			
Perceiving Nonverbal Messages.			
Persuading			
Reporting of Information......			
Describing Feelings..........			
Interviewing................			
Editing			
Additional Skills:			

SKILLS INVENTORY

SKILL	✔	RATING 1–3	EVIDENCE
RESEARCH AND PLANNING: The search for specific knowledge and the ability to conceptualize the future and create the process for getting there.			RESEARCH AND PLANNING:
Forecast, Predict			
Creative Ideas			
Identification of Problem			
Imagine Alternatives..........			
Identification of Resources.....			
Gathering Information			
Problem Solving			
Goal Setting.................			
Extracting Important Information			
Defining Needs..............			
Analyzing			
Developing Evaluation Strategy			
Additional Skills:			

SKILLS INVENTORY

SKILL	✔	RATING 1–3	EVIDENCE
HUMAN RELATIONS SKILLS: The use of interpersonal skills for resolving conflict, relating to and helping people.			HUMAN RELATIONS SKILLS:
Developing Rapport			
Sensitivity			
Listening..................			
Conveying Feelings			
Drawing Out People			
Motivating			
Sharing Credit			
Counseling			
Cooperation			
Delegating with Respect for Others			
Representing Others			
Perception of Feelings, Situations			
Assertiveness			
Additional Skills:			

SKILLS INVENTORY

SKILL	✔	RATING 1–3	EVIDENCE
ORGANIZATION, MANAGEMENT, or LEADERSHIP: The ability to supervise, direct, and guide individuals and groups in the completion of tasks and fulfillment of goals.			ORGANIZATION, MANAGEMENT, or LEADERSHIP:
Initiating New Ideas..........			
Handling Details.............			
Decision Making			
Coordinating Tasks...........			
Managing Groups			
Delegating Responsibility			
Teaching....................			
Coaching....................			
Counseling			
Promoting Change............			
Selling an Idea...............			
Decision Making with Others ..			
Conflict Management			
Additional Skills:			

RECOMMENDED READINGS

Useful manuals on style and method abound. *The Modern Researcher* (4th ed.), (New York: Harcourt Brace Jovanovich, Inc., 1985) by Jacques Barzun and Henry Graff ranks as a classic. Other references include *Historians and the Living Past, The Theory and Practice of Historical Study* (Arlington Heights, IL: AHM Publishing Corporation, 1978) by Allan J. Lichtman and Valerie French; *Understanding History, A Primer of Historical Method*

(2nd ed.), (New York: Alfred A. Knopf, 1969) by Louis Gottschalk; *A Guide to Historical Method* (Homewood, IL: The Dorsey Press, 1969) edited by Robert Jones Shafer; *Creative History* (2nd ed.), (Philadelphia: J.B. Lippincott Company, 1973) by W.T.K. Nugent; *Historian's Handbook, A Key to the Study and Writing of History* (2nd ed.), (Boston: Houghton Mifflin Co., 1964) by Wood Gray and others; and *Studying History, An Introduction to Methods and Structure* (3rd ed.), (Washington, D.C.: American Historical Association, 1985) by Paul L. Ward. Writing techniques and other issues come under consideration in *A Manual of Style* (12th ed.), (The University of Chicago Press, 1969); *Harbrace College Handbook* (9th ed.), (New York: Harcourt Brace Jovanovich, Inc., 1982) by John C. Hodges and Mary E. Whitten; and *The Elements of Style* (New York: Macmillan, Inc., 1959) by William Strunk, Jr. and E.B. White.

7

PROFESSIONAL HISTORY IN RECENT TIMES

The professionalization of history in the middle of the nineteenth century produced profound changes in focus and direction and inspired fierce struggles over purposes and methods. As practitioners labored initially to carry out the agenda of "scientific" history, they sought to obtain truths about the past through the utilization of assiduous research and stringent techniques. Their successors placed less credibility in simple, naive faiths. Some doubted the accessibility of unquestionable verities and urged instead the acquisition of useful insights, emphasizing the importance of the relative angle from which historians viewed their subjects. Others, seeking more inclusive and reliable methods, hoped to possess deeper levels of understanding by extending the bounds of inquiry beyond the political, military, and diplomatic activities of male elites and through the application of the social sciences. The ensuing endeavors had mixed results. Though historical studies became more precise and thorough, the preoccupations of professionalization also resulted in isolation and fragmentation. As a consequence of rampant specialization, common varieties of historical understanding broke down, and in their place, pluralistic interpretations gave testimony to the disparities of human experience.

The outcomes reflect the very nature of recent times. In an important work entitled *Historiography, Ancient, Medieval and Modern*, the historian Ernst Breisach argued that four characteristics particularly affected historical thinking in the years after 1870. These consisted of the impacts of science, the process of industrialization, the emergence of mass culture, and the perception of a global world. To elaborate, science, often regarded as the model of all knowledge, set profound epistemological problems before historians, while the disruptions of the industrial revolution destroyed traditional conceptions of unity and continuity. In similar fashion, the emergence of mass movements, the concentration of populations in urban regions, and radical political appeals among "the people" directed attention to life among the common folk, while the triumph of the West over much of the remainder of the globe altered traditional ideas of world history. In combination, the effects shattered uniform pictures of the past and resulted in great divergences.

During the closing years of the nineteenth century, the so-called "hard" sciences acquired immense prestige for their supposed capacities to produce demonstrable claims about reality and to develop reliable predictive devices. Because such presumed certainties possessed irresistible appeal, students of human affairs, including historians, looked toward them as examples to emulate. But the effort collided with enormous conceptual difficulties, chief among them the distinction between the general and the particular. Though of course not always the case, the sciences most characteristically focused on the former and history on the latter. If historians should substitute a concern for the common and the repeatable for the singular and the unique, they might lose identity altogether. If universal principles and "covering-law" theories should become the objects of their search, then history would have to undergo a transformation in the image of physics or chemistry or else would cease to exist as a legitimate kind of investigation with a just claim to knowledge.

The epistemological controversies already reviewed in an earlier chapter set "positivists" against "idealists" and historians among themselves. Though innovators in response invoked the social sciences in the quest for methodological reassurance and scientific exactitude, traditionalists resisted the allure of arguments by analogy, insisting that in their field the term "science" possessed a more restricted meaning. For them, methodological rigor required that they strip away unwarranted theological and philosophical assumptions, accumulate the evidence, and then scrutinize it coldly, analytically, and objectively without preconceptions. Once understood in appropriate detail and with the necessary degree of detachment, history, as observed by the English historian J. B. Bury in 1903, would amount to a science, "nothing more, nothing less."

Other historians, spurred on by the harsh, unrelenting circumstances of the industrial age, arrived at new understandings of the "true" forces shaping history. For them, in an era of dramatic and jolting economic change, abstract and idealistic conceptions of the universe, no matter whether defined in supernatural or metaphysical or rationalistic terms, held less reality than the press of material conditions. According to the materialistic view, the actual world consisted of economic struggle to obtain a livelihood, and the pursuit of wealth functioned as a prime determinant of human behavior. Karl Marx set forth his positions on these subjects in the middle of the nineteenth century, and henceforth, subsequent renditions, both of the Marxist and non-Marxist varieties, directed attention toward the organization of the productive system and the means by which people earned their livings. Such pursuits turned into economic history.

The burgeoning urban masses also compelled notice and contributed to new kinds of social history. They focused on common, ordinary people. Though never completely anonymous in historical narratives, such persons, typically powerless, inconspicuous, and caught up in the cycles of birth, labor, and death, normally took subordinate places to the dominant, ruling elites who controlled politics, diplomacy, and war. Indeed, the very historical records employed by historians—political, diplomatic, and military—restricted access to the lesser folk. They showed up mainly in the announcements of birth and death. Though Marxist analysis emphasizing the other classes required some readjustments, most nineteenth-century historians who wrote about "the people" had in mind collectivities of human beings who together found their stories incorporated into the histories of their nations. In the twentieth century, in contrast, historians turned their investigations to the development of group identities, particularly workers, peasants, racial and ethnic types, women, and families. Within such categories, the main historical actors oftentimes could not find their stories within the traditional, national histories, and they demanded new ones.

Finally, the Western ascendency over much of the rest of the world required revisions of traditional conceptions of universal history. Almost always ethnocentric, the various renditions, whether providential or secular, Christian or Marxist, characteristically took Europe as the model of historical development for all of humanity and displayed little sensitivity to other peoples in other cultures. Oswald Spengler, of course, ranked as a notable exception. In such interpretive schemes, differences appeared as deviances, and European standards became the norms, indeed, the very hallmarks of civilization itself. In the nineteenth century, imperial triumphs in Asia, Africa, and the Middle East impressed Europeans not only as manifestations of an ordained, civilizing mission in the barbarous regions but as irrefutable signs of their own intrinsic superiority, defined in Darwinian terms, over the lesser breeds. In the United States, an ideological counterpart affirmed the doctrines of manifest destiny. Under the impact of such deceits, universal history became merely a projection of the European (or North American) experience, part and parcel of the progressive thrust moving the inhabitants of the globe ever onward. In recent times, certain sociological and economic theories of "modernization" based on Western models, for example, W. W. Rostow's *The Stages of Economic Growth, A Non-Communist Manifesto* (1960), incorporated vestiges of the theme, but for the most part in much of the rest of the world, such ideas lost credibility after 1945, when the Europeans lost their dominance and their empires. As a consequence, Western scholars with a serious interest in universal or world history had to start again, seeking out new categories and analyses by which to make comprehensible the histories of non-European peoples of their own terms.

At the end of the nineteenth century, European historians took a special interest in the continuities of historical experience. Focusing mainly on the institutional development of constitutional and legal systems, scholars explored questions of national identity and indulged in arcane discussions over the origins. The issues held certain patriotic implications. In Germany and France, medievalists such as Heinrich von Sybel and Fustel de Coulanges debated whether the primary influences shaping society had Frankish or Roman antecedents. In England too, investigations revolved around constitutional history and the implication of the Whig interpretation, pointing to the central importance of the extension of liberty. Whether the English nation had its roots in ancient Anglo-Saxon or European practices became a matter of vivid significance.

In the United States similarly, a preoccupation with institutional history expressed national pride and a sense of national identity. In the middle of the nineteenth century, George Bancroft and other amateur, "literary" historians had rejoiced in the nation's birth and growth, seeing in it indications of divine favor. Such conceptions of organic unity and supernatural direction broke down after the Civil War, in part the result of industrialization, urbanization, immigration, and class strife. As a consequence, a new generation of university-trained historians in the 1880s and 1890s sought to harness more rigorous and reliable means of historical analysis. The creation of the American Historical Association in 1884 testified to their commitment to professionalism.

Inspired by visions of "scientific" history, American historians incorporated into their work initially the European concern for institutional growth. Early renditions traced the lineage of American liberty and democracy back to the Germanic heritage. Professor Herbert Baxter Adams preached this gospel in his famous seminar at Johns Hopkins University in Baltimore, locating the "germ" of later developments in the misty Teutonic forests. The idea, rudimentary in form, could not withstand scientific scrutiny and dissolved under pressure from various sources, including the so-called "imperial school" of colonial history. Scholars such as Herbert L. Osgood, George Louis Beer, and Charles McLean Andrews traced the roots not to remote German villages but to the English legacies and also to the actual effects of the colonial experience in the New World.

The social dislocations of the age and the ensuing reform impetus called forth new approaches in academic scholarship, particularly in the areas of law, philosophy, social science, and history. Seeking to promote progress through reform and concious planning, historians during the Progressive era in the United States developed the so-called "New History." Characteristically skeptical in tone and iconoclastic in approach, the "New" historians especially regarded lofty, public statements of high-blown and idealistic purpose with suspicion, perceiving them as subterfuge for material or selfish interests.

An early version, really a precursor, appeared in the works of Frederick Jackson Turner. In the year 1893, he read before the American Historical Association a paper on the significance of the frontier in American history. His seminal idea, the so-called frontier thesis, retained great influence within the historical profession for over half a century and profoundly shifted the focus of study in United States history. First, Turner repudiated the views of Herbert Baxter Adams, his teacher, and all notions that American institutions had germinated in the Teutonic woods. Turner saw liberty, democracy, and individualism as homegrown traits, the products of the frontier experience. Second, he shifted attention away from the eastern seaboard and New England toward the West, where successive acts of settlement had engendered the laudable qualities of which he spoke. Oftentimes elusive and poetic in formulation, Turner's sparse writings never made clear whether the author understood the frontier as a place, a process, or a state of mind. Nevertheless, the Turner thesis captivated historians because, in a single stroke, it accounted for American uniqueness and put the ordinary people, the settlers, at the center of things.

Another among the "New" historians who shaped the consciousness of his times, Charles Austin Beard developed an economic interpretation of United States history. More indebted intellectually to James Madison and *The Federalist Papers* than to Karl Marx, Beard, a committed activist and reformer, first directed his attention toward the founding fathers. In 1913, the publication of *An Economic Interpretation of the Constitution* set forth daring, heretical views. Emphasizing the impact of class conflicts, Beard disparaged the supposed effects of altruistic impulses and providential designs by underscoring the complex interplay of material concerns. Indeed, he characterized the clash as a contest between a popular party based on paper money and agrarian concerns and a conservative party founded upon urban, financial, mercantile, and personal property interests. For Beard, the adoption of the Constitution, a triumph for the latter, really amounted to a kind of counterrevolution against the alleged, democratic excesses under the Articles of Confederation. In a later work entitled *The Rise of American Civilization*, first published in 1934, Beard and his wife Mary gave special prominence to economic causes and argued the case that United States history consistently had pitted the interests of the business classes against those of the people.

Especially poignant during the decade of the Great Depression, such themes figured prominently in the Progressive school of historiography in the United States. During the years of war and upheaval, Progressive historians retained hopes for the future and expressed a cherished regard for democratic traditions out of concern for maintaining them in a violent, industrial age. Through the mechanisms of social science, rationality, cooperation, and planning, they hoped to retain the best of the past in the years to come and also sought to facilitate through their scholarship the very cause of reform.

Their works centered on conflict in the American experience, defined either in class or sectional terms. In a study of the origins of the American Revolution, *A History of Political Parties in the Province of New York, 1768–1776* (1909), Carl L. Becker depicted a struggle not only against English authority but against the dominant, colonial oligarchy, possessing a monopolistic hold on power and wealth. As Becker noted, it became a question not only of home rule but also of who should rule at home. The sectional emphasis, to an extent inspired by Frederick Jackson Turner, gave a conspicuous role to competition among regions with distinctive geographic, economic, and social characteristics. Handled in this way, United States history turned into a story of incessant rivalry among the East, the South, and the West.

Though Turner concentrated on the frontier regions, and an entire generation of Western historians debated the merits of his thesis, other scholars found inspiration elsewhere. The works of Vernon Parrington, Perry Miller, and Samuel Eliot Morison illuminated New England traditions, especially the Puritan heritage. The South with its "peculiar institution," slavery, provoked sympathetic investigations by William A. Dunning and a cluster of historians around him. The Civil War figured as a central event for Progressive historians. Although Beard treated this climactic struggle in economic terms, a second American Revolution bringing about the unchallenged dominance of the industrial classes, his revisionist successors, notably James G. Randall in *The Civil War and Reconstruction* (1937), explained it as the consequence of inexplicable human folly, the catastrophic work of "a blundering generation." The questions of causation and inevitability, whether the war was avoidable or unavoidable, took on fundamental significance.

Oddly, Progressive historians displayed little interest in scrutinizing the course of race relations in the United States. Indeed, to an extent, some writings incorporated racist assumptions. Ulrich B. Phillips understood slavery as a way of coping with an "inferior" race. Accordingly, the writing of black history fell to black historians, notably W. E. B. Du Bois and Carter G. Woodson, both educated at Harvard. During long careers, they engaged in political and intellectual struggles to enhance the status of black people and to construct more positive identities. Du Bois, for example, participated in founding both the National Association for the Advancement of Colored People (NAACP) and *The Journal of Negro History*.

In spite of the foregoing, lamentable lapse, Progressive historians tried to write useful history in order to elucidate the nature of contemporary social issues. Their pragmatism set some of them at odds with the standards of "scientific" history and raised questions about the attainability of the objective truth. For Carl Becker and Charles Beard, "That Noble Dream," in Beard's words, would always elude fulfillment. For them, public utility as much as accuracy should guide the construction of historical narratives. When Beard spoke dubiously of objectivity before the American Historical Association in 1935, he raised a storm over his dangerous flirtation with the doctrine of relativism. If the criterion of objectivity would not hold, then historians could never assure themselves that they knew anything at all.

Doubt and uncertainty assailed historians all over the world during the decade of the Great Depression, although in some countries totalitarian renditions provided the illusion of safe haven. In Italy, fascism exalted the leadership principle and identified Benito Mussolini as the manifestation of the national will but failed to impose this version of reality upon important Italian intellectuals. Benedetto Croce, a philosopher and historian, dissented from fascist orthodoxy, and Antonio Gramsci, a Marxist sociologist, paid for his opposition with his life. In Germany similarly, the doctrine of National Socialism and the racism attached to it never prevailed within the universities, although to be sure the purges of Jewish professors sacrificed some of that country's best minds. Nazi historiography venerated German warriors, military power, and the authority of the state but never achieved a unified vision of the past. The legacies of the Nazi debacle left future generations of German historians with the obligation to wrestle with the problems of guilt and responsibility.

In the Soviet Union meanwhile, Marxism became the state ideology. Though traditionally much influenced by German conceptions of rigor and objectivity, Russian historians of bourgeois origin became suspect during the Stalin era when, indeed, the dictator intervened in academic disputes in order to ensure "correct" interpretations. As a consequence, Bolshevik doctrines became unassailable, and the Russian Revolution an unquestioned anticipation of the future. According to the orthodox view, the Soviet Union would assume leadership in bringing about the fulfillment of Marxist-Leninist prophecies and the establishment of world communism. Though for a time during the Second World War, termed in Russian "the Great Patriotic War," the desperation of the life-and-death struggle called forth an emphasis on more traditional nationalism, the reappearance of the hard-line Stalinist version after the peace reaffirmed the hegemony of the official ideology not only within the Soviet Union but over "the peoples' republics" in Eastern Europe as well.

Elsewhere in the world, Marxian scholars, removed from Stalin's authority, struggled with the intellectual puzzles of reconciling fact with theory and thought with practice. Among the members of the so-called "Frankfurt school" in Germany during the 1930s, critical thinkers repudiated simple-minded varieties of "vulgar" Marxism, featuring an undue emphasis only on economics, and urged a fuller appreciation of the diversity of influences shaping human behavior. Erich Fromm, for example, sought some form of amalgamation with psychoanalysis. During the Hitler years, the dispersion of the Frankfurt school into other countries introduced the constituents (for example Herbert Marcuse) to the rigors of Anglo-Saxon empiricism and at the same time invigorated more conventional thinking outside Germany with exposure to sophisticated Marxian analysis.

French scholars also participated in the effort to obtain more comprehensive forms of understanding of the human past. Led by Marc Bloch and Lucien Febvre, the founders in 1929 of the influential journal, *Annales d'histoire économique et sociale*, the historians making up the so-called "*Annales* school" rebelled against the prevailing forms of academic history. Rejecting the narrow emphasis on politics, war, and diplomacy, what Febvre disparaged as *histoire événementielle*, that is, event-oriented history, the *Annales* group strove to grasp more totally and fully the whole dimensions of human reality. As the historiographer Ernst Breisach explained, these French scholars envisioned a new, more complete history, inclusive of all aspects of human life. To achieve the aim, they aspired to the development of a larger repertoire of investigatory techniques and called for extensive cooperation with "comrades and brothers" across the entire span of social and human sciences. As Febvre expressed the goal, "Down with all barriers and labels! At the frontiers, astride the frontiers, with one foot on each side, that is where the historians has to work, freely, usefully."[1]

The tenets of the *Annales* group achieved triumph among French historians after 1945, testifying in part to a sense of disenchantment with the practices and values of the old France, a principal loser in the Second World War. Resuming publication under a new title, *Annales: Economies, Sociétés, Civilisation,* *Annaliste* scholars resumed their quest for more total history. Surprisingly, although their work had implications for historical theory and methodology, they wrote little on such subjects. During the war, Marc Bloch, the famed medievalist, then a hunted resistance fighter later executed by the Nazis, composed a book, *The Historian's Craft,* a modest methodological survey but hardly a manifesto. By and large, *Annales* scholars left their theoretical and methodological assumptions implicitly in their other writings.

Two traits figured conspicuously. First, *Annaliste* scholars typically built their analyses around a conception of collective consciousness. Termed *mentalité,* this phenomenon focused attention on the mental and psychological characteristics of groups of people at specified times and places and thus moved historians beyond constrictive and sometimes myopic concerns with mere individuals. According to this approach, the collectivity counted most in the formulation of explanations leading to total history. Second, in similar fashion, *Annaliste* historians employed a notion of the *longue durée,* the long duration. Actually a conception of time, this term depicted the structural continuities intruding upon the course of historical change. The *longue durée,* among other things, comprised the land, the sea, the climate, and the vegetation. These conditions affirmed stabilizing influences over the conduct of human affairs and transpired at a slower pace or rhythm than the transitory events of politics of politics, war, and diplomacy. They determined, moreover, the manner of life.

The best-known work by an *Annaliste* historian, Fernand Braudel's *The Mediterranean World in the Age of Philip II* first appeared in 1949 with a second edition in 1966. It set forth ambitious aims, seeking to encompass the whole, the totality of life in the region, focusing upon the uniformities in the political, social, economic, intellectual, and geographic realms. Braudel's concern for the continuities of life, the overarching structures of time, and the languid pace of change, and also his efforts to develop conceptual unities, obtained many impressive results, but according to his critics seldom achieved the desired level of integration. The sheer magnitude of the task militated against the fulfillment of his good intentions but nevertheless inspired successors in the *Annales* school, such as Emmanuel Le Roy Ladurie, who wrote on peasants and rural civilization; Philippe Ariés, who studied childhood and death; and Michael Foucault, who concentrated on madness. Such innovative works kept alive the ideal of total history.

In the United States, the impact of the war and its aftermath shook the foundations of Progressive historiography. Indeed, successive cataclysmic horrors instilled new appreciations for indeterminacy and uncertainty, some things of which the Europeans had a firmer grasp. Simple verities lost appeal. As Reinhold Niebuhr explained in *The Irony of American History* (1952), even the best of intentions could go awry, resulting in error and terrible excess. The Progressive school's commitment to plain dualities, the juxtaposition of selfish business interests against the democratic will of the people, held less credibility, and as a consequence, new ornate and more problematic conceptions of historical realities came to the fore.

The publication in 1945 of Arthur Schlesinger's important work, *The Age of Jackson*, constituted a kind of monument to the attainments of Progressive historiography. Though subtly put, it developed a characteristic argument favoring the continuity of reform in American politics from Thomas Jefferson to Andrew Jackson to Franklin Roosevelt, but with an intriguing twist. Rather than find the impetus to reform among agrarians and Westerners (the Turner view) Schlesinger located it among urban workers and Eastern liberals. Schlesinger's book set off a flurry of research, writing, and reevaluation. Although some scholars still operated within the context of the Progressive tradition, notably Merrill Jenson in his study of the United States during the Articles of Confederation, *The New Nation* (1950), and Eric F. Goldman's work on reform movements since 1870, *Rendezvous with Destiny* (1952), the dominant, historiographic tendency in the immediate future put a quite different construction upon the American past.

Richard Hofstadter's *The American Political Tradition and the Men Who Made It* (1948) heralded the shift. Widely read and immensely influential, it lamented the current "lack of confidence in the American future" and "the rudderless state of

American liberalism." Fearful that the experiences of history had not prepared the American people to respond creatively in the years to come, Hofstadter found the fault in a traditionally opportunistic style of political leadership, "a democracy of cupidity," and an absence of truly fundamental differences separating liberals from conservatives. Since the two groups habitually agreed upon essentials, for example, the legitimacy of profit incentives and private property, the Progressive emphasis upon conflict in United States history obscured the basic reality. Quite the contrary, according to Hofstadter, the main contenders for power affirmed a consensus in support of the business and capitalism.

Hofstadter sounded the theme so characteristic of historical thinking in the 1950s, and other scholars soon elaborated upon it from different points of view. Perhaps part of a collective effort to obtain stable moorings in a sea of global change, scholars agreed that homogeneity more than disparity marked the American experience and that consensus held more explanatory power than conflict. But they differed over the reasons. In *The Liberal Tradition in America* (1955), Louis Hartz, a political scientist, explained the uniformities by arguing that Lockean assumptions about political reality bound together the American people. Born directly into the modern world, unlike the Europeans, they had no feudal structures and titled nobility from which true conservatism could emerge. In *The Americans: The Colonial Experience* (1958), Daniel Boorstin attributed the cause to the stark, ineluctable, and practical necessities of surviving in the New World, and David M. Potter, in *People of Plenty, Economic Abundance and the American Character* (1954), found it rooted in material conditions.

Historians in postwar America also reacted against the relativistic implications of Progressive historiography and sought to develop more reliable methodologies, capable of producing verifiable results. Characteristically, these scholars looked toward the social sciences and quantification techniques. In an important work entitled *The Concept of Jacksonian Democracy* (1961), Lee Benson employed statistical analysis in order to show that Arthur Schlesinger's categories in *The Age of Jackson* would not hold up in New York state, and hence, the description of the party and class alignments lacked veracity. Though the prospect of obtaining precise measurements in such matters has held allure ever since the ancient Greeks, the drive to use numbers turned into a kind of crusade and gained momentum, in spite of Schlesinger's put-down. As he phrased it, "As a humanist, I am bound to reply . . . that almost all important questions are important precisely because they are not susceptible to quantitative answers."[2]

Unsurprisingly, quantification and the use of social scientific methodology first took hold in economic history. As a discipline, economics yielded early to systemic analysis and econometrics, both of which viewed reality as consisting of coherent systems, the parts of which combined in linkages one with the other according to strict patterns. In any such arrangement, changes in any one part necessarily resulted in transformations of the whole. Moreover, mathematical models theoretically could provide the means to analyze, to assess, and to predict the consequences and the magnitudes of change, thereby, in effect, obtaining the standards of the natural sciences.

In the United States after 1945, the so-called "New Economic history," or "cliometrics," aimed at achieving such methodological sophistication. No longer would economic historians describe and explain specific kinds of events. Instead, they would attain greater universality by scrutinizing categories of events, aggregates, and group behavior. Among the foremost practitioners, Robert W. Fogel, in a book entitled *Railroads and American Economic Growth: Essays in Econometric History* (1964), affirmed a controversial claim. Seeking to test the thesis that railroads held central importance in the economic development of the United States, Fogel developed carefully framed models to control variables and to explore a counterfactual hypothesis. Would the absence of railroads have made any difference? Remarkably, he concluded, no, it would not, since Americans would have found altenative kinds of transportation. In a second book, *Time on the Cross: The Economics of American Negro Slavery* (1974), Fogel and his colleague Stanley L. Engerman stirred up a profound dispute by depicting "the peculiar institution," contrary to prevailing views, as efficient, profitable, and conducive to a standard of living, defined exclusively in material terms, at least as high as that of Northern workers. Critics responded with denunciations of this work as a methodological monstrosity, reminiscent of antiabolitionist propaganda before the Civil War.

If quantification could obtain such levels of meaning in economic history, why not also in other areas? To an extent following the lead of the *Annales* school in France, various historians proposed to employ such means in the pursuit of more total forms of understanding. Statistical analysis and counting became the orders of the day in the "new" political and social histories, and the tasks of collecting, storing, and processing the data became ever easier because of modern computers. Accordingly, the utilization of collective biographies within influential groups and the exhaustive examination of election returns affirmed demonstrable conclusions about political behavior. Presumably, the same held true in investigations centering on the formation of elites, the distribution of property ownership, the degree of social mobility, and the demographic structure. In the case of the latter, amazingly, through the proper use of projections and approximations based on population statistics and parish records, historians could arrive at reasonably exact rates of marriage, fertility, and mortality. Often unimpressed by such attainments, more traditional historians sometimes regarded them as mere shows of manipulative virtuosity, dazzling performances in which the means of investigation too frequently turned into ends in themselves. The critics wondered whether an understanding of history really benefited much. In response, William O. Aydelotte defended quantification in history with the modest claim that it provided "a means of verifying general statements."[3]

Another proposed use of social science advised the application of psychoanalytical theory. Though less precise than mathematical formulations, this endeavor also strove for deeper forms of understanding. Focusing on individual behavior, it normally found the wellspring in the human psyche, particularly in the conflict between inner drives and outer constraints. For orthodox Freudians, the experiences of infancy and the relations with parents and siblings held special importance in shaping the future adult. For example, the much lamented and unfortunate psychoanalytical biography of Woodrow Wilson by William C. Bullitt and Sigmund Freud, *Thomas Woodrow Wilson, A Psy-*

chological Study (1967) attributed the man's deficiencies, notably his need to fail, to the incapacities of the boy and his inability to satisfy the demands of an insatiable father.

Psychohistory, the amalgamation of psychoanalytical theory with history, acquired growing influence in the United States after 1945, in part because of diminished faith in reason and progress and also because of an influx of European experts. Their views won over influential American historians (for example, William Langer) and encouraged the utilization of Freud's findings in order to plumb the depths of the psyche. According to psychohistorians, a greater awareness of the role of unreason in human behavior would result in better history.

The ensuing scholarly efforts have had mixed results. Since effective psychohistory compels expertise in two demanding disciplines, the requirements sometimes have exceeded the capabilities of individual practitioners and have produced inadvertently comic consequences. In more capable hands, in contrast, psychoanalytical techniques employing theories of character formation have achieved notable results. Examples would include Erik H. Erikson, *Young Man Luther* (1958), David H. Donald, *Charles Sumner and the Coming of the Civil War* (1960), and Fawn M. Brodie, *Richard Nixon, The Shaping of His Character* (1981). Yet even such expert works raised compelling questions, persuading dubious observers to resist some of the inflated claims of champions, such as Lloyd de Mause, who saw in psychohistory an emerging science of human motivation. Critics mounted attacks from two main directions. On the one hand, they assailed the tendency to reduce history to individual biography, thereby endorsing new versions of the old-fashioned and discredited Great Man (or Human) theory. On the other, they skeptically appraised the possibility of verifying any claims set forth. Indeed, according to this view, psychohistory entailed the most blatant of empathetic leaps into the heads of historical actors and allowed for no believable methods of proof at all. Groundless and theoretical, it set forth no means of verification.

Though the precepts of social-scientific history remained a source of inspiration for many historians, others in the 1960s and after harkened back in a sense to the traditions of Progressive historiography. Seeking to put political commitments back into their narratives, they preferred contemporary relevance over scientific detachment and sought, much as the Progressives, to elucidate current issues through historical inquiries. The conduct of foreign affairs, race relations, women's issues, and the use and abuse of political and economic power became objects of compelling concern. Obviously reflecting current cares over the Vietnam War, the civil rights movement, the drive for women's liberation, and the Watergate affair, many historians making up the so-called "New Left" reworked the usual versions of United States history. Characteristically, they sought more radical perspectives, for example, by incorporating viewpoints from "the underside," representative of the lower classes, the racial and ethnic minorities, and women. Often harshly critical in tone, New Left historiography usually depicted the established system as rapacious, exploitative, and racist in character, and often attributed such excesses to the conditions of economic capitalism. Though aspiring to avoid dogmatism, most New Left historians maintained that Marxism had applicability to the American experience.

Such revisionism challenged more orthodox historiography on many points and raised firestorms of dispute. An early statement, William Appleman William's *The Tragedy of American Diplomacy* (1959), repudiated the usual interpretation, emphasizing the good if oftentimes futile intentions of the United States, by depicting the country's foreign policy as the product of capitalist requirements. According to Williams, the Marxian emphasis on markets, trade, and profits held true. Consequently, a group of his students and associates at the University of Wisconsin elaborated upon such themes, constituting the so-called "Wisconsin school." Walter LaFeber's *The New Empire; An Interpretation of American Expansion, 1860–1898* (1963), for example, set forth an economic interpretation of United States expansion around the turn of the century, and Lloyd C. Gardner's *Architects of Illusion, Men and Ideas in American Foreign Policy, 1941–1949* (1970) did similarly for the Cold War era.

Such New Left renditions and other efforts inspired by them impelled searching reevaluations of United States history and veered off in many directions. The economic system, seen as a corporate entity designed, controlled, and maintained by rul ing elites, entailed new forms of understanding, reaching also to the structure of class, race, and gender relations. Corollary efforts illuminated the histories of ethnic and racial minorities, notably blacks and Chicanos, and also resulted in a tide of publications on women. The outcomes, provocative and revealing, usefully showed the feasibility of doing history from "the underside" but also promoted a degree of conceptual fragmentation.

The more traditional narratives, focusing on the activities of white, male elites, had presumed commonalities of experience. Supposedly such stories spoke for everyone. The newer histories, in contrast, raised questions and insisted upon distinctions. For example, when the civil rights movement in the 1950s and 1960s riveted public attention on race relations in the United States, the initial historical works by and large dealt with the issue as it appeared to whites. This approach never satisfied the aggrieved minorities. The black perspective later appeared in important works, such as John W. Blassingame's *The Slave Community, Plantation Life in the Antebellum South* (1972) and Eugene D. Genovese's, *Roll, Jordan, Roll, The World the Slaves Made* (1974). These books tried to comprehend the institution of slavery from the inside, that is, by showing how it appeared to and affected black people and how they coped with the circumstances.

Similarly, the writing of women's history posed profound conceptual problems. Women in the feminist movement properly complained that textbook history particularly had rendered them invisible through exclusion. They just did not show up. Although to be sure a few historians had written about women, the publications consisted of a particular sort, focusing either on "great" women, such as Joan of Arc or Queen Elizabeth, or political activists in the reform and suffrage movements, such as Jane Addams. Critics observed correctly that women had to act like men in order to merit attention.

The new scholarship often conveyed a feminist emphasis and insisted upon putting ordinary women into the historical narratives. Neither "great" nor necessarily politically active, they included middle-class women, immigrants, ethnics, and all racial groups. Yet, the mere organization of the material presented puzzles, because traditionally historians had derived the very categories in which they thought from the world of males. For example, chronological schemes based on kingly reigns or presidential administrations, economic cycles, or warfare had nothing intrinsically to do with women. The format of the conventional textbook, in short, had scant relevance. How then to incorporate the history of women?

This question pressed heavily upon scholars and allowed for no single solution. In a groundbreaking book entitled *Woman's Proper Place, A History of Changing Ideals and Practices, 1870 to the Present* (1978), Sheila Rothman suggested one response. For her, the study of women in history required an approach consisting of three parts. First, historians must comprehend the roles and responsibilities assigned to women in any given period. Next, the scholars must determine the degree to which various women of different classes and races actually adhered to those demarcations. Last, they must carefully observe the process of change, the shifting definitions of women's proper place over time, and the degrees to which the various categories of women actually complied with them.

Such conceptual and methodological conundrums have confounded practitioners throughout the discipline of history and to an extent have erected barriers against communication. Highly specialized in their forms of expertise, historians sometimes lament that they have difficulty understanding and appreciating the work of colleagues in other fields. The cacophony produced by clashing stratagems and rival interests represents, among other things, a demographic change among the scholars in the profession. No longer dominated by the scions of aristocratic, East-coast families, the historians include among their ranks persons of diverse origins whose objects of study testify to their diversity. Together with different angles of vision, they participate in the collective enterprise of writing history.

Although particularities and divergences characterize the craft of history in the present day, one thing seems reasonably clear. History no longer sets forth common stories which presumably speak for the identity and experience of all readers. For many consumers of history, the narratives centering on the activities of white male elites no longer provide either satisfaction or stimulation. We no longer possess a past commonly agreed upon. Indeed, to the contrary, we have a multiplicity of versions competing for attention and emphasizing alternatively elites or nonelites, men or women, whites or nonwhites, and no good way of reconciling them. Though the disparities and the incoherencies create terrible predicaments for historians who prize orderliness in their stories, such conditions also aptly express the confusions of the world and the experiences of different people in it. If historians can agree on little else, at least they should rejoice in the knowledge that they have many true stories to tell about the past.

RECOMMENDED READINGS

These titles represent a large body of scholarship. *Historians at Work*, volumes 3 and 4, *Niebuhr to Maitland*, and *Dilthy to Hofstadter* (New York: Harper & Row, Publishers, Inc., 1975), by Peter Gay, Victor G. Wexler, and Gerald J. Cavanaugh, shows the shift from the nineteenth to the twentieth century. *Historiography, Ancient, Medieval & Modern* (The University of Chicago Press, 1983), by Ernst Breisach, spells out the details and contains a splendid bibliography. Other books delineate trends and current tendencies, among them, *The Past Before Us, Contemporary Historical Writing in the United States* (Ithaca, NY: Cornell University Press, 1980), edited by Michael Kammen; *Historical Studies Today* (New York: W.W. Norton & Co., Inc., 1972), edited by Felix Gilbert and Stephen R. Graubard; and *History, The Development of Historical Studies in the United States* (Englewood Cliffs, NJ: Prentice-Hall, Inc., 1965), by John Higham, Leonard Krieger, and Felix Gilbert. Marc Bloch's classic discussion of methodology is contained in *The Historian's Craft*, translated by Peter Putnam (New York: Vintage Books, 1953).

Works on specific historians include Richard Hofstadter, *The Progressive Historians, Turner, Beard, Parrington* (New York: Alfred A. Knopf, Inc., 1969); Lee Benson, *Turner and Beard, American Historical Writing Reconsidered* (New York: The Free Press, 1960); Harvey Wish (ed.), *American Historians, A Selection* (New York: Oxford University Press, 1962); and Earl E. Thorpe, *Black Historians, A Critique* (New York: William Morrow & Co.,

Inc., 1971). John Barker, *The Super-Historians, Makers of Our Past* (New York: Charles Scribner's Sons, 1982), contains an essay on W.E.B. Du Bois. The views of Charles A. Beard and Carl Becker are set forth in "That Noble Dream," *American Historical Review* 41 (Oct. 1935); and *Everyman His Own Historian, Essays on History and Politics* (New York: F. S. Crofts & Co., 1935).

Insights into the ongoing debates over research methods are obtained in *Historical Analysis, Contemporary Approaches to Clio's Craft* (New York: John Wiley & Sons, 1978), edited by Richard E. Beringer; *Toward the Scientific Study of History, Selected Essays* (New York: J.B. Lippincott Company, 1972), by Lee Benson; *American Historical Explanations, A Strategy for Grounded Inquiry* (Homewood, IL: The Dorsey Press, 1973), by Gene Wise; *A Behavioral Approach to Historical Analysis of Historical Data* (Homewood, IL: The Dorsey Press, 1969), edited by D.K. Rowney and J.Q. Graham; *Quantification in History* (Reading, MA: Addison-Wesley, Publishing Co., Inc., 1971), by William O. Aydelotte; and *Which Road to the Past? Two Views of History* (New Haven, CT: Yale University Press, 1983), by G.R. Elton and Robert W. Fogel. Jacques Barzun, *Clio and the Doctors, Psycho-History, Quanto-History & History* (The University of Chicago Press, 1974), and David E. Stannard, *Shrinking History, On Freud and the Failure of Psychohistory* (New York: Oxford University Press, 1980), raise probing questions.

Articles on historiography sometimes appear in *The American Historical Review*. Recent issues contain "Annaliste Paradigm? The Geohistorical Structure of Fernand Braudel," 86 (Feb. 1981), by Samuel Kinser; "The Contribution of Women to Modern Historiography in Great Britain, France, and the United States, 1750–1940," 89 (June 1984), by Bonnie Smith; and "Guilt, Redemption, and Writing German History," 88 (Feb. 1983), by Theodore S. Hamerow. The volume devoted to *The American Historical Association: The First Hundred Years, 1884–1984*, 89 (Oct. 1984), features essays on "Historical Consciousness in Nineteenth-Century America," by Dorothy Ross; "Beyond Consensus: Richard Hofstadter and American Historiography," by Daniel Joseph Singal; and "J. Franklin Jameson, Carter G. Woodson, and the Foundations of Black Historiography," by August Meier and Elliott Rudwick.

NOTES

Chapter 1

1. Paul K. Conkin and Roland N. Stromberg, *The Hertitage and Challenge of History* (New York: Dodd, Mead & Company, 1971), p. 131.

Chapter 2

1. Quoted in Herbert Butterfield, *The Origins of History* (New York: Basic Books, Inc., Publishers, 1981), pp. 76–77.
2. Michael Grant, *The Ancient Historians* (New York: Charles Scribner's Sons, 1970), pp. 10–11.
3. Herodotus, *The Histories*, trans. Aubrey de Sélincourt (Baltimore: Penguin Books, 1954), p. 13.
4. Quoted in Grant, *Ancient Historians*, p. 78; Thucydides, *The Peloponnesian War*, trans. John H. Finley, Jr. (New York: The Modern Library, 1951), p. 14.
5. Quoted in Grant, *Ancient Historians*, p. 78.

Chapter 3

1. B.A. Haddock, *An Introduction to Historical Thought* (London: Edward Arnold, Publishers, Ltd., 1980), p. 45.
2. R.G. Collingwood, *The Idea of History* (New York: Oxford University Press, 1956), p. 76.
3. Peter Gay and Victor G. Wexler, eds., *Historians at Work, Volume 2: Valla to Gibbon* (New York: Harper & Row, Publishers, Inc., 1972), p. 84.
4. Collingwood, *Idea of History*, p. 77.
5. From "The Age of Louis XIV" in Gay and Wexler, eds., *Historians at Work*, 2: 285.
6. In Gay and Wexler, eds., *Historians at Work*, 2:357. From Edward Gibbon, *The Decline and Fall of the Roman Empire*, ab. Moses Hadas (New York: Capricorn Books, 1962), p. 131.
7. Quoted in Gay and Wexler, eds., *Historians at Work*, Volume 3: *Niebuhr to Maitland* (New York: Harper & Row, Publishers, 1975), p. 89, Thomas Babington Macaulay, *The History of England, From the Accession of James II*, 5 vols. (New York: A.L. Burt Co., Publishers, n.d.), 1:2–3.
8. Collingwood, *Idea of History*, pp. 90–91.
9. From the "Introduction" to Hegel's *The Philosophy of History* in Monroe G. Beardsley, ed., *The European Philosophers from Descartes to Nietzsche* (New York: The Modern Library, 1960), p. 553.
10. Quoted in Gay and Wexler, eds., *Historians at Work*, 3:16; Leopold von Ranke, *The Theory and Practice of History*, ed. Georg G. Iggers and Konrad von Moltke (New York: The Bobbs-Merrill Co., Inc., 1973), p. 137.
11. Quoted in Bruce Mazlish, *The Riddle of History, The Great Speculators from Vico to Freud* (New York: Minerva Press, 1966), p. 227.

Chapter 4

1. Quoted in Bruce Mazlish, *The Riddle of History, The Great Speculators from Vico to Freud* (New York: Minerva Press, 1966), p. 38. See also *The New Science of Giambattista*

Vico, Revised Translation of the Third Edition (1744), tr. Thomas Goddard Bergin and Max Harold Fisch (Ithaca, N.Y.: Cornell University Press, 1968), pp. 335–351.

2. In Ronald H. Nash, ed., *Ideas of History, Volume 1: Speculative Approaches to History* (New York: E. P. Dutton & Co., Inc., 1969), p. 61.

3. Nash, ed., *Ideas of History*, 2:54.

4. Quoted in Mazlish, *Riddle*, p. 336; Oswald Spengler, *The Decline of the West*, abr. Helmut Werner (New York: The Modern Library, 1962), pp. 73–74.

5. Mazlish, *Riddle*, p. 320.

6. Arnold J. Toynbee, *A Study of History*, abr. D.C. Somervell (New York: Oxford University Press, 1957), 2:357–358.

7. Sigmund Freud, *An Outline of Psychoanalysis*, tr. James Strachey (New York: W.W. Norton & Co., 1949), pp. 16–17.

8. Reinhold Niebuhr, *Faith and History* (New York: Charles Scribner's Sons, 1949), p. 7. Copyright renewed © 1977 Ursula M. Niebuhr. Reprinted with permission of Charles Scribner's Sons.

Chapter 5

1. The foregoing comes from R.G. Collingwood, *The Idea of History* (New York: Oxford University Press, 1956), pp. 204–215.

2. Collingwood, *Idea of History*, p. 223.

3. Carl G. Hempel, "The Function of General Laws in History," in *Theories of History*, ed. Patrick Gardiner (New York: The Free Press, 1959), pp. 344, 348–349. First published in *The Journal of Philosophy* 39, no. 2 (15 January 1942), 36.

4. Ronald H. Nash, ed., *Ideas of History, Volume 2: The Critical Philosophy of History* (New York: E.P. Dutton & Co., 1969), p. 185.

5. Paul K. Conkin and Roland N. Stromberg, *The Heritage and Challenge of History* (New York: Dodd, Mead & Company, 1971), p. 211.

6. Isaiah Berlin, *The Hedgehog and the Fox, An Essay on Tolstoy's View of History* (New York: Mentor, 1957), pp. 7–8.

Chapter 7

1. Ernst Breisach, *Historiography, Ancient, Medieval, and Modern* (Chicago: University of Chicago Press, 1983), pp. 370–371.

2. Quoted in Breisach, *Historiography*, p. 340.

3. Quoted in Breisach, *Historiography*, p. 340.

INDEX

A

Acton, Lord, 2, 17
Adams, Henry, 40
Adams, Herbert Baxter, 110, 111
American Historical Association, 110, 114
Ancient historiography, 3, 10, 11, 12, 49
Andrews, Charles McLean, 111
"*Annales* school," 115–17, 120
Annals, 20, 21
Ariés, Philippe, 177
Aristotle, 80
Augustine, St., 19, 20, 23, 30, 36, 49, 57, 66, 68
Aydelotte, William O., 120

B

Bacon, Francis, 70
Bancroft, George, 40, 44, 110
Barnes, Robert, 30
Beard, Charles A., 83, 84, 111–12, 113, 114
Beard, Mary, 112
Becker, Carl L., 83, 112, 113
Bede, Venerable, 22–23
Beer, George Louis, 111
Benson, Lee, 118
Berlin, Sir Isaiah, 84–85
Blassingame, John W., 123
Bloch, Marc, 31, 115, 116
Bodin, Jean, 31
Boorstin, Daniel, 118
Boronius, Caesar, 30
Bossuet, Bishop Jacques Bénigne, 32
Bradford, William, 33
Braudel, Fernand, 116
Breisach, Ernst, 107, 115
Brodie, Fawn M., 121
Buckle, Henry Thomas, 72, 74

Bullitt, William C., 120
Bury, J. B., 108

C

Caesar, Julius, 17
Camden, William, 32
Careers, 98–104
Cause and effect, 5–6, 7
Christian historiography, 10, 17, 18, 19, 20–24, 30, 49–50
Chronicles, 20, 21, 24
Collingwood, Robin G., 6, 32, 36–37, 41, 74–77, 79
Comte, August, 72–74
Condorcet, Marie Jean Antoine, 52, 53
Conkin, Paul K., 3, 83
Coulanges, Fustel de, 110
"Critical" history, 4, 13, 27
Croce, Benedetto, 74–75

D

De Mause, Lloyd, 121
Dependency theory, 57–58
Descartes, René, 2, 37, 38, 70–71
Dialectic, 42, 51
Dialectical materialism, 45–46, 54–57
Dilthey, Wilhelm, 74
Donald, David H., 121
Dray, William, 79
Du Bois, W. E. B., 113
Dunning, William A., 113

E

Eliade, Mircea, 49
Engels, Friedrich, 54
Engerman, Stanley, 119